ΣΔ A/D CONVERSION FOR SIGNAL CONDITIONING

THE INTERNATIONAL SERIES IN ENGINEERING AND COMPUTER SCIENCE

ANALOG CIRCUITS AND SIGNAL PROCESSING
Consulting Editor*: Mohammed Ismail. *Ohio State University

Related Titles:

WIDE-BAND WIDTH HIGH-DYNAMIC RANGE D/A CONVERTERS
Doris, Konstantinos, ven Roermund, Arthur, Leenaerts, Domine
Vol. 871, ISBN: 0-387-30415-0
METHODOLOGY FOR THE DIGITAL CALIBRATION OF ANALOG CIRCUITS AND SYSTEMS: WITH CASE STUDIES
Pastre, Marc, Kayal, Maher
Vol. 870, ISBN: 1-4020-4252-3
HIGH-SPEED PHOTODIODES IN STANDARD CMOS TECHNOLOGY
Radovanovic, Sasa, Annema, Anne-Johan, Nauta, Bram
Vol. 869, ISBN: 0-387-28591-1
LOW-POWER LOW-VOLTAGE SIGMA-DELTA MODULATORS IN NANOMETER CMOS
Yao, Libin, Steyaert, Michiel, Sansen, Willy
Vol. 868, ISBN: 1-4020-4139-X
DESIGN OF VERY HIGH-FREQUENCY MULTIRATE SWITCHED-CAPACITOR CIRCUITS
U, Seng Pan, Martins, Rui Paulo, Epifânio da Franca, José
Vol. 867, ISBN: 0-387-26121-4
DYNAMIC CHARACTERISATION OF ANALOGUE-TO-DIGITAL CONVERTERS
Dallet, Dominique; Machado da Silva, José (Eds.)
Vol. 860, ISBN: 0-387-25902-3
ANALOG DESIGN ESSENTIALS
Sansen, Willy
Vol. 859, ISBN: 0-387-25746-2
DESIGN OF WIRELESS AUTONOMOUS DATALOGGER IC'S
Claes and Sansen
Vol. 854, ISBN: 1-4020-3208-0
MATCHING PROPERTIES OF DEEP SUB-MICRON MOS TRANSISTORS
Croon, Sansen, Maes
Vol. 851, ISBN: 0-387-24314-3
LNA-ESD CO-DESIGN FOR FULLY INTEGRATED CMOS WIRELESS RECEIVERS
Leroux and Steyaert
Vol. 843, ISBN: 1-4020-3190-4
SYSTEMATIC MODELING AND ANALYSIS OF TELECOM FRONTENDS AND THEIR BUILDING BLOCKS
Vanassche, Gielen, Sansen
Vol. 842, ISBN: 1-4020-3173-4
LOW-POWER DEEP SUB-MICRON CMOS LOGIC SUB-THRESHOLD CURRENT REDUCTION
van der Meer, van Staveren, van Roermund
Vol. 841, ISBN: 1-4020-2848-2
WIDEBAND LOW NOISE AMPLIFIERS EXPLOITING THERMAL NOISE CANCELLATION
Bruccoleri, Klumperink, Nauta
Vol. 840, ISBN: 1-4020-3187-4
CMOS PLL SYNTHESIZERS: ANALYSIS AND DESIGN
Shu, Keliu, Sánchez-Sinencio, Edgar
Vol. 783, ISBN: 0-387-23668-6
SYSTEMATIC DESIGN OF SIGMA-DELTA ANALOG-TO-DIGITAL CONVERTERS
Bajdechi and Huijsing
Vol. 768, ISBN: 1-4020-7945-1
OPERATIONAL AMPLIFIER SPEED AND ACCURACY IMPROVEMENT
Ivanov and Filanovsky
Vol. 763, ISBN: 1-4020-7772-6
STATIC AND DYNAMIC PERFORMANCE LIMITATIONS FOR HIGH SPEED D/A CONVERTERS
van den Bosch, Steyaert and Sansen
Vol. 761, ISBN: 1-4020-7761-0
DESIGN AND ANALYSIS OF HIGH EFFICIENCY LINE DRIVERS FOR Xdsl
Piessens and Steyaert
Vol. 759, ISBN: 1-4020-7727-0
LOW POWER ANALOG CMOS FOR CARDIAC PACEMAKERS
Silveira and Flandre
Vol. 758, ISBN: 1-4020-7719-X
MIXED-SIGNAL LAYOUT GENERATION CONCEPTS
Lin, van Roermund, Leenaerts
Vol. 751, ISBN: 1-4020-7598-7

ΣΔ A/D CONVERSION FOR SIGNAL CONDITIONING

by

Kathleen Philips

Philips Research Laboratories,
Eindhoven, The Netherlands

and

Arthur H. M. van Roermund

Eindhoven University of Technology,
The Netherlands

Springer

A C.I.P. Catalogue record for this book is available from the Library of Congress.

ISBN-10 1-4020-4679-0 (HB)
ISBN-13 978-1-4020-4679-7 (HB)
ISBN-10 1-4020-4680-4 (e-book)
ISBN-13 978-1-4020-4680-3 (e-book)

Published by Springer,
P.O. Box 17, 3300 AA Dordrecht, The Netherlands.

www.springer.com

Printed on acid-free paper

Printed in the Netherlands.

Contents

List of symbols and abbreviations

Symbol	Description	Unit
BW	bandwidth of the wanted signal	Hz
C	capacitance	F
c	linearized quantizer gain for a ΣΔ ADC	
$C_O(t)$	matched filter	
CCh	conditioning channel	
d	linearized DAC gain for a ΣΔ ADC	
dB_{FS}	ratio (in decibels) of the signal power compared to the full-scale power of a digital data format	dB_{FS}
DR	dynamic range; i.e. ratio of the maximum signal power to the minimum detectable signal power in the same bandwidth	
DR_{dB}	DR expressed in decibels	dB
DSP	Digital Signal Processing	
F	implementation factor, i.e. the ratio of $FOM(ADC)$ over $FOM(reference\ ADC)$	
f_a	transition frequency from a high-order to a first-order slope of a transfer function	Hz
FFB-ADC	filtering-feedback ΣΔ ADC	
f_{ug}	unity-gain bandwidth	Hz
f_s	Nyquist sample rate	
f_{sw}	average switching frequency; i.e. the inverse of the average number of $0-1$ transitions in a digital data sequence	Hz
IM_3	ratio of the amplitude of the third-order intermodulation component, to the amplitude of the fundamental signals. The fundamental signals both are applied at half of the full-scale level.	
$IM_{3,dB}$	IM_3 expressed in decibels	dB_c
k	Boltzmann's constant equaling $1.38 \times 10^{-23}\,J/K$	J/K
L	order of the loop filter of a ΣΔ ADC	
m	over-sampling factor	
MASH	multi-stage noise shaping	
N	number of bits of a quantizer	

Symbol	Description	Unit
NTF	noise transfer function	
p	ratio between the bandwidth of the entire signal (including the wanted and interferer channels) and the bandwidth of the wanted signal only	
P	power consumption	W
q	ratio of the amplitude of an overall signal, consisting of both wanted and interferer components, to that of the wanted signal only	
$SINAD$	signal/noise-and-distortion ratio; i.e. ratio of the power of the wanted signal to the sum of the noise power and the distortion power. All are integrated over the channel bandwidth and are present simultaneously	
$SINAD_{dB}$	SINAD expressed in decibels	dB
SNR	signal/noise ratio; i.e. ratio of the power of the wanted signal to the noise power. Both are integrated over the channel bandwidth and are present simultaneously	
SNR_{dB}	SNR expressed in decibels	dB
STF	signal transfer function	
T	absolute temperature	K
$\hat{v}_{\mathrm{IN}}$	instantaneous amplitude of a wanted signal	V
$\hat{v}_{\mathrm{MIN}}$	minimum amplitude of a wanted signal	V
$\hat{v}_{\mathrm{MAX}}$	maximum amplitude of a wanted signal	V
v_n	short-hand notation for the rms value of an input-referred, equivalent noise source, measured over the wanted channel only	V
V_{DD}	positive supply voltage	V
V_{SS}	negative supply voltage	V

Chapter 1

Introduction

1.1 Background

Moore's Law predicts a decrease by a factor of two in the feature size of CMOS technology every three years and has been valid for years. It implies a doubling of the operation speed and a four times higher transistor count per unit of area, every three years. The combination leads to an eight times higher processing capability per unit of area. This on-going miniaturization allows the integration of complex electronic systems with millions of transistors (Very-Large-Scale-Integration) and enables the integration of electronic systems.

An electronic system

A generic picture of an integrated electronic system is shown in fig. 1.1. The heart of the system is the signal processing core. This core supports a wide variety of functions, such as customization and programmability of multiple applications, channel coding, the definition of the user interface, etc. These functions are enabled by DSP, a controller CPU and various blocks of memory. In advanced ICs these blocks provide (almost) all signal processing and usually dominate in the overall power and area consumption of integrated systems. The huge data rates involved, require high-speed busses for communication between these blocks. A power-management unit fuels the system by providing the appropriate supply voltages and currents.

Communication with the physical world is realized by a chain of mixed-signal, analog and RF-circuitry. This chain acts as a "signal-conditioning channel". It translates physical signals into binary representations or vice versa. This channel also comprises amplification, filtering and possibly frequency translation. It is the challenge of present day mixed-signal and RF design to integrate the signal-conditioning channel at a low power consumption and a high performance level.

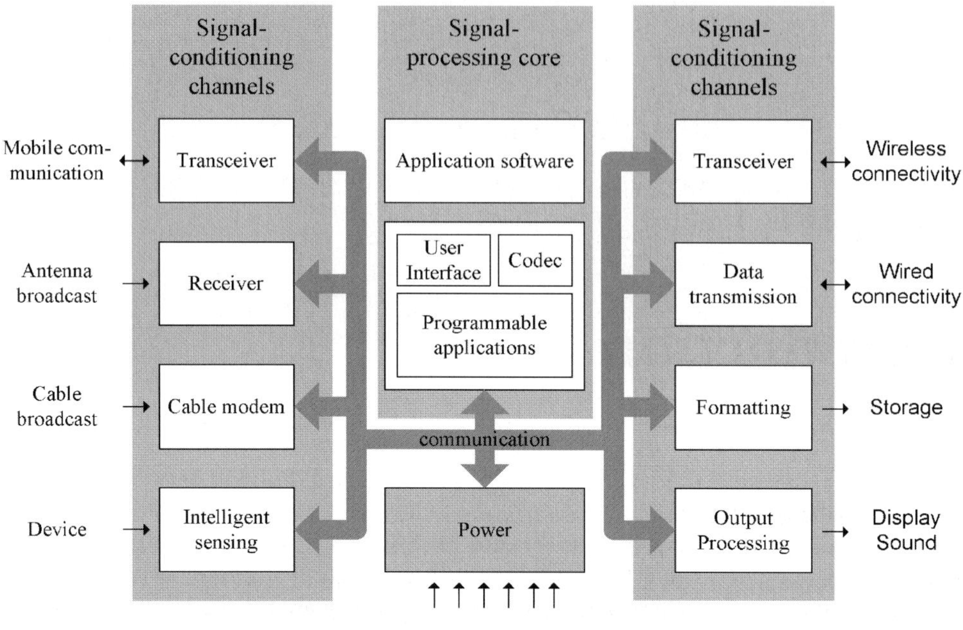

Figure 1.1: *Electronic system including "signal-processing" units and some example "signal-conditioning" channels for communication to the physical world [1]*

Digital signal processing

More and more digital systems and standards have been conceived: DECT has replaced analog cordless systems, CD has overtaken the market of analog audio and digital cameras are conquering conventional photography. These new standards offer higher quality and more features thanks to the digital signal processing. The digital solution is moreover easily scalable to new technologies and changing systems.

Clearly, *in the signal-processing arena "full-digital" is becoming a fact for the majority of systems. A similar evolution is happening to the signal-conditioning channels shown in fig. 1.1 although the feasibility and economics of "highly-digitized signal conditioning" have not yet been proven.*

Digitization of signal conditioning

Fig. 1.2 depicts a conventional (i.e. dominantly analog) implementation of an A/D and a D/A type of signal-conditioning channel. In fact, this could be the block diagram detailing any of the signal-conditioning channels introduced in fig. 1.1. These channels select the wanted signal in the time and the frequency domain, amplify the wanted signal, suppress interferer signals and noise, perform A/D or D/A conversion and de- or encoding of the signal.

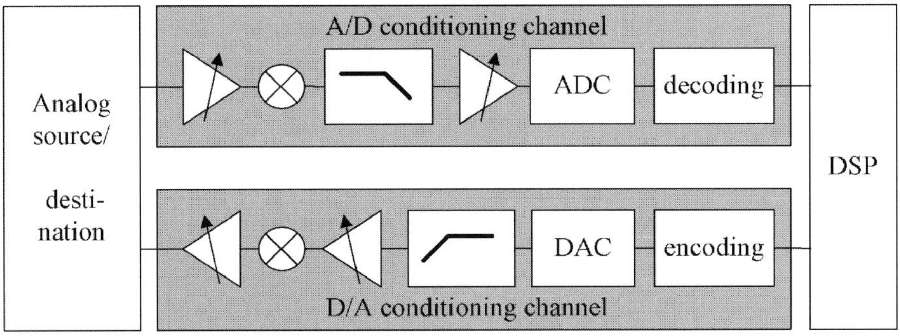

Figure 1.2: *Block schematic of conventional A/D and D/A conditioning channel*

The signal conditioning happens in either the analog or the digital domain[1]. Predominant analog signal conditioning relaxes the bandwidth and the dynamic range requirements of all the following blocks. Predominant digital conditioning reduces the number of analog blocks needed, while the implementation of the digital blocks benefits from Moore's Law. The requirements imposed on the data converters, however, become substantially stricter. As the ADC and DAC move towards the antenna a much higher sample rate and significantly higher resolution and linearity are required (see example on page 15).

While digital processing seems to come for free in advanced CMOS technologies, any performance increase of analog circuitry leads to higher power consumption. Moreover, migration of an analog circuit to a new technology generation may imply a power increase, even at constant performance requirements [3], [4]. Although some transistor parameters have improved in advanced technologies, the negative effect of the decreasing supply voltage has over-compensated this for generations beyond 0.25μm-CMOS [5]. Therefore, new circuit techniques need to be developed in order to bridge this performance gap. Evolution in analog circuit techniques is, however, very slow in comparison with the revolution that has taken place in the complexity of digital processing and algorithms.

Obviously, *digitization of the signal-conditioning channel in an advanced CMOS technology, imposes a substantial burden on the analog circuits and on the data converters especially. All-digital signal conditioning is therefore not necessarily the best option in view of the overall optimization of the signal-conditioning channel.*

Digitization of inter-die interfaces

The signal-conditioning channel can be integrated on a single die or often it extends over multiple dies (in order to take advantage of a dedicated technology or for reasons of standardization of the interface). In the latter case, digitization of the channel may lead to a

[1] In this book, it is assumed that de- and encoding of the signal occurs in the digital domain. This is obvious for digital communication schemes but is also becoming the de-facto implementation for analog systems. A good example are FM radio receivers in which analog demodulation -although adequate and low-cost- is substituted by digital demodulation [2] featuring all the previously mentioned advantages.

(signal-)conditioning channel =

analog/digital filtering + variable gain/word-length scaling + data conversion

signal conditioning

Figure 1.3: *Nomenclature with respect to the signal-conditioning channel*

new inter-die interface: as the ADC or DAC shifts towards the antenna a previously analog interface may be replaced by a digital interface. Hence, the cost of an inter-die interface will strongly depend on the degree of digitization of the signal-conditioning channel, and must be included in the overall optimization of the cost/performance ratio of the channel.

1.2 Scope

This book studies the digitization of the signal-conditioning channel. In particular, it focuses on the consequences of digitization on the power consumption of the channel in relation to the performance target. The target "performance" is evaluated in terms of noise, distortion and bandwidth. In generic terms, the aim of this book is to *improve the power/performance relation of the conditioning channel by balancing analog and digital signal conditioning*. This is pursued while striving for a highly-digitized solution. Some limitations on the scope of the text are briefly motivated next.

Baseband A/D conditioning channels

The text focuses on signal conditioning in an A/D type of channel, operating at baseband. The key circuits in the analysis are filters, variable-gain amplifiers and data converters. At present, digitization of this (part of the) signal-conditioning channel -though very challenging still- is becoming feasible for various systems. On the contrary, power-efficient digitization at the IF or RF frequency can be considered as a next -but further-off- step. In addition, the de- and encoding of the signal are left out because the digitization of these blocks has become a reality already.

The nomenclature as defined in fig. 1.3 will be used. "Conditioning channel" is used as a shorthand notation for "signal-conditioning channel". As explained above, it only refers to the conditioning actions identified in fig. 1.3. Moreover, the term "signal conditioning" only refers to filtering and variable gain or word-length scaling, without the data conversion. In chapter 6, this functionality is integrated into a $\Sigma\Delta$ ADC. Then, the terminology of "a conditioning $\Sigma\Delta$ ADC" is used.

Continuous-time single-bit sigma-delta conversion

The choice for sigma-delta data conversion is motivated by the evolution of CMOS technology and the system need for low power data conversion (see chapters 2 and 3) and by the potential of sigma-delta converters for digitizing the conditioning channel in a power efficient way. The latter argument is demonstrated throughout the book.

This choice does limit the analysis to channels with a "narrow" bandwidth. At present, $\Sigma\Delta$ ADCs with a signal bandwidth up to 40MHz have been reported [6] in CMOS. For this range of bandwidths $\Sigma\Delta$ converters enable low-power, high-performance conditioning channels.

This is especially true in case a continuous-time loop filter is used as this alleviates the requirements on preceding anti-aliasing filtering. In addition, most continuous-time implementations have a better power/performance ratio than their switched-capacitor counterparts often due to lower bandwidth requirements on the filter stages [7].

Single-bit quantization provides high-linearity. This is a major specification on the ADC in case analog conditioning -limiting bandwidth and *DR* of the input signal- is traded for digital conditioning.

Motivated by the above promise of an attractive cost/performance ratio the text concentrates on continuous-time, single-bit $\Sigma\Delta$ -ADCs (see also 3.2.1).

CMOS technology

The choice for a baseband mixed-signal channel justifies a further narrowing of the scope to CMOS technology only. While CMOS is gaining ground in many application areas, it is certainly doing so in the field of analog and mixed-signal baseband design. This follows from the number of scientific publications in this area.

Power consumption as cost parameter

The optimization of the signal-conditioning channel is performed in a single cost dimension, being power consumption. This is certainly a viable choice for portable applications aiming at long stand-by times and small and light-weight battery packs. In general, low power consumption can be an important asset in view of limited heat sinking capabilities of packages, in view of area required by fans, etc.

The analysis aims at calculating the relation between the current consumption and the performance requirements. The maximum supply voltage is assumed to be dictated by the technology choice.

Performance parameters

In view of the comparison of various architectures for the conditioning channel a limited set of performance parameters needs to be identified. These parameters need to represent a fundamental specification on a generic signal-conditioning channel, influence the balancing of analog and digital conditioning and influence the power consumption. Based on Shannon's theory on the capacity of a generic communication channel a meaningful set of parameters is derived in section 2.2. These parameters relate to signal bandwidth, signal amplitude, noise power and distortion. The associated nomenclature is discussed in section 2.5.

For now it is mentioned that only (white) thermal circuit noise is taken into account. Other noise sources like flicker noise and shot noise only occur in a limited frequency band and therefore are less generic.

Only differential circuits are considered such that third-order distortion dominates. Differential operation is preferable in a mixed-signal environment anyway. Furthermore it is assumed that all circuits operate under weakly non-linear conditions which implies that the response at the nth harmonic is only determined by the nth order non-linearity.

1.3 Outline

The book starts with a study of a generic signal-conditioning channel in chapter 2. Applying Shannon's theory, the choice of the performance parameters is further motivated. It is explained how system evolution and technology advances affect the conditioning channel and digitization is identified as a key challenge. In addition, some nomenclature is introduced.

Chapter 3 presents an overview of state-of-the-art in ΣΔ A/D converter design and motivates the assets of ΣΔ converters for digitization of the conditioning channel. For completeness, we briefly touch upon limitations on the application of ΣΔ converters.

In chapter 4, power/performance relations for the building blocks of the conditioning channel -i.e. for a major class of analog circuits, for the ΣΔ ADC and for the decimation filter- are derived. This leads to conclusions on how to proceed in view of power-efficient digitization. Further on, these results are used to compare conditioning channels, with a varying degree of digitization, in terms of their power/performance balance.

In chapter 5, we study a full-analog and a full-digital conditioning channel. These represent the two extremes in terms of digitization and are compared with respect to power consumption.

Chapter 6 introduces the concept of "conditioning ΣΔ ADCs". Instead of having analog conditioning, in front of the ADC, or performing the conditioning in the digital domain, it is integrated into the ΣΔ loop. This concept is enabled by the fact that ΣΔ ADCs are largely immune to interferers. The analysis of the interferer immunity and of the limitations thereon, is a key topic of this chapter. In addition, various ΣΔ topologies are evaluated in this perspective and a "filtering-feedback ΣΔ ADC" -explicitly designed for interferer immunity- is presented. Again, the power/performance balance of the various solutions is assessed as well.

Often, the signal-conditioning channel extends over multiple dies. In that case, digitization of the conditioning channel, may lead to digitization of the inter-die interface as well. This is the topic of chapter 7.

Chapters 8 and 9, present design examples as an illustration of the theory of chapters 5 and 6 respectively. A dual-mode receiver for FM/AM radio is considered in chapter 8. In FM mode, the signal conditioning is highly analog. In AM mode, it is highly digitized using multi-channel A/D conversion. In chapter 9, three implementations of a "conditioning ΣΔ ADC" for use in a Bluetooth receiver are discussed. The first design is attractive for systems requiring a high *SNR* for the digital processing. The other two designs enable power-efficient digitization of applications requiring less *SNR*. Especially the filtering-feedback $\Sigma\Delta$ ADC is very promising because of its flexibility and an inherently low-power architecture.

Finally, in chapter 10, general conclusions are summarized.

Chapter 2

The signal conditioning channel

This chapter describes the conditioning channel in more detail. Shannon's theory on a generic communication channel is applied to motivate the selection of a key set of performance parameters. Next, the basic functionality of the conditioning channel is presented.

While the functional requirements on the conditioning channel are fixed, it is shown that the performance targets increase due to system evolution and the digitization demand. At the same time, the opportunities arising from circuit innovation or technology evolution are limited. As such, this chapter reveals a key challenge on the implementation of the conditioning channel in today's playing field.

Finally, some nomenclature is introduced.

2.1 Generic communication channel

A generic communication channel is represented in fig. 2.1. A message encoded in a signal is transported through a channel to a destination. Various noise sources add to the wanted signal but should not corrupt the message. The capacity C of a communication channel has been calculated by Shannon as a function of the bandwidth (*BW*) and the signal/noise ratio (*SNR*) of the channel. In case of a Gaussian signal source and Gaussian noise sources the following well-known formula results:

$$\text{channel capacity} = BW \cdot \log_2(1 + SNR) \tag{2.1}$$

In order not to loose information from source to destination, a match is needed between, on one hand, the *BW* and the *SNR* of the channel and, on the other hand, the capacity required by the signal source and the destination. The required match is expressed by equation 2.1.

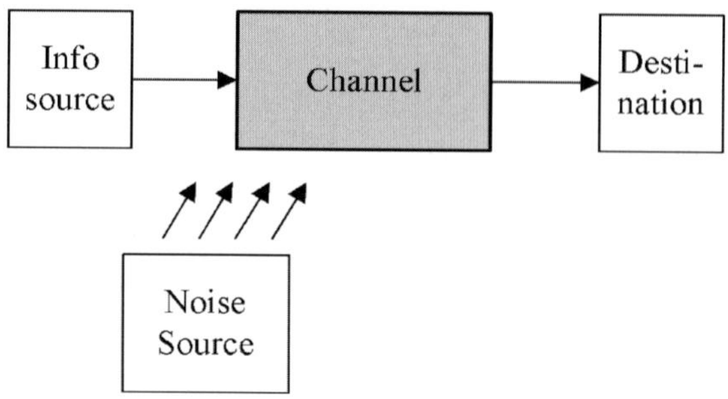

Figure 2.1: *A generic communication channel as described by Shannon*

2.2 Performance parameters

The Shannon model is used to derive a set of performance parameters to characterize the channel. Equation 2.1 shows that the *BW* and *SNR* of the channel are fundamental to its performance. Of course, Shannon's theory assumes a Gaussian signal source and Gaussian noise sources. It has been derived for a simplified model of a communication channel. Still, it gives an indication of the key performance parameters for an arbitrary, non-ideal communication channel.

In a practical application, the signal source is not Gaussian. In fact, the channel performance is often specified for a sinusoidal input signal. In addition, non-Gaussian noise and interference sources such as distortion, aliasing due to a sampling operation or interference from unwanted signals (for example substrate bounce) will be present. These "noise" sources may further limit the channel capacity below what is predicted by eq. 2.1. This limitation may be reversible for some of these "noise"sources; for instance distortion can be reduced by means of calibration. For other noise sources -like for instance interference- this becomes more difficult. In this text, only thermal noise and distortion are included in the calculation. Compared to substrate bounce, aliasing, etc., they represent a more generic specification on the implementation and very often noise and distortion are dominant anyway.

Fig. 2.2 gives a schematic summary of this discussion: based on Shannon's theory, and considering practical linearity constraints, *signal bandwidth, signal amplitude, noise power and distortion* are identified as the key parameters for characterizing the performance of a generic conditioning channel. The exact metrics and the associated symbols are discussed in section 2.5. First, the link between the Shannon channel and the signal conditioning in an electronic system is made.

2.3 Conventional conditioning channels

The communication channel that is studied in this book is the signal-conditioning channel in a highly digitized electronic system (fig. 1.1). In such a system the actual processing of the information in the signal is performed in the digital domain. The signal-conditioning

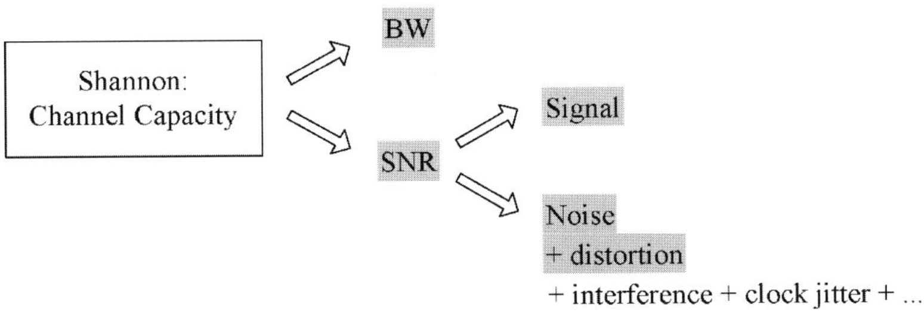

Figure 2.2: *Interference and noise sources in a communication channel*

channels provide communication from the physical world to the signal processing core of the system and vice versa. A conventional (i.e. dominantly analog) implementation of both types of conditioning channels has been presented in fig. 1.2.

The *functionality* of the A/D conditioning channel includes frequency translation from the carrier frequency to baseband. The channel needs to discriminate the wanted signal from interferers and noise, involving selection in the frequency and in the time domain. The dynamic range of the input signal must be reduced to the resolution required for digital processing and, finally, A/D conversion, including sampling and quantization, is performed.

In addition, Shannon's theory gives a boundary condition on correct communication through the channel. In fact, it expresses a *performance target* for the channel. *BW* and *SNR* of the conditioning channel must be high enough to provide the required channel capacity (equation 2.1). Since the generic conditioning channel doesn't match the ideal Shannon channel this performance target should be interpreted in a broader sense: *BW* and *DR* of the channel must match that of the incoming signal -with a sufficient margin- while the *SINAD* and sample rate must be sufficient to accommodate the output data rate required for the digital signal processor.

The functionality of the D/A conditioning channel is -to a large extent- complementary to that of the A/D channel. It includes D/A conversion including reconstruction and interpolation, suppression of spurious components around multiples of the sample frequency, scaling of the amplitude of the signal to the required output level and frequency translation from baseband to the carrier frequency. A similar performance target as in the A/D channel needs to be achieved.

2.4 Evolution

Fig. 2.3 depicts the playing field for implementing the conditioning channel. System level requirements set the performance target on the conditioning channel. From the technology side, opportunities and limitations on the silicon implementation of the channel arise. Fig. 2.3 also shows the degrees-of-freedom when implementing a conditioning channel; i.e. the definition of the architecture, the design of analog circuits and the digital signal

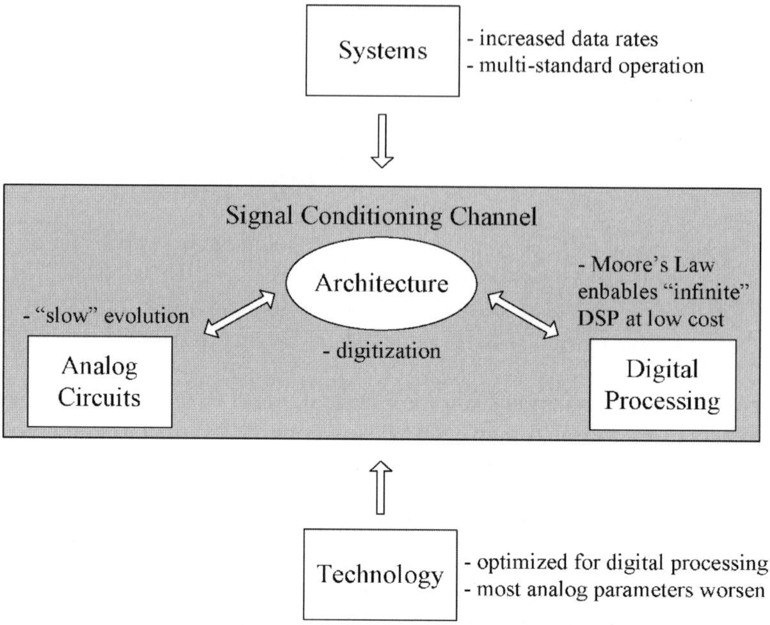

Figure 2.3: *Playing field and degrees-of-freedom for implementing the conditioning channel*

processing. Below, the various elements in this picture are discussed only briefly in order to stick to the defined scope. *The discussion does indicate a future direction for the implementation of the conditioning channel.*

2.4.1 Technology advances

CMOS technology is optimized for digital processing and strives for more computational capability per unit of area. Recently, also some attention is devoted to low-power digital operation and to leakage problems.

On the contrary, for generations beyond the $0.25\mu m$-node, technology evolution became unfavorable for analog circuit design [8]. Here, the threats of deep-submicron seem more numerous than the opportunities. Some advantages are available, though:

- higher g_m/C, slightly higher g_m/I, slightly better matching, etc.

Especially the bandwidth improvement can be exploited. However, many analog parameters worsen:

- V_{DD} drops, the output impedance of transistors decreases, the bulk effect (and the associated non-linearity) becomes more important, thermal noise increases slightly[1]

[1] Note that the *Noise Figure* of the transistor improves because its transconductance increases more than the thermal noise.

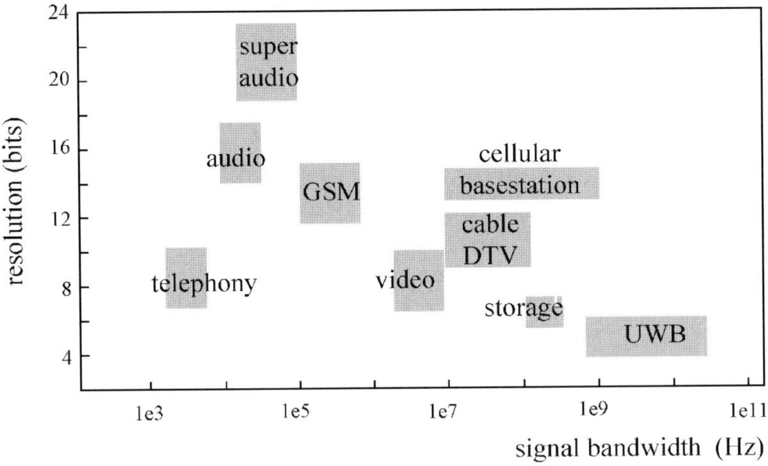

Figure 2.4: *System demands in terms of signal bandwidth and resolution*

[9], etc.

It becomes very difficult to obtain high DC gain or linear operation for the MOS devices. For a more thorough discussion, the reader is referred to [8], [10], [11], [12], [3], [13], etc.

Note that *the characteristics of advanced CMOS technologies match the requirements for ΣΔ ADC design.* ΣΔ modulation exchanges resolution in amplitude for resolution in the time-domain. As such, most analog sub-blocks of the ΣΔ ADC are not critical with respect to distortion, noise or matching. On the contrary, the circuit bandwidth must increase because of the over-sampling. This is elaborated in section 3.2.2.

2.4.2 System demands

Fig. 2.4 plots a number of systems as a function of the resolution and bandwidth requirement for the A/D converter. Established systems for speech, digital audio or analog television are at the less demanding side of the diagram. More recent systems, such as the super-audio standard, high-speed storage, GSM, etc., need more resolution and/or bandwidth. Future systems, a.o. Ultra-Wide-Band data communication, etc., move further to the extremes of the axes.

In fact, this graph demonstrates the higher bit rate demand of new systems. This increase is due to various factors:

- more information needs to be exchanged;
- higher quality is demanded by the customer;
- more encoding "overhead" is needed because of a much denser use of the available spectrum, space and time slots

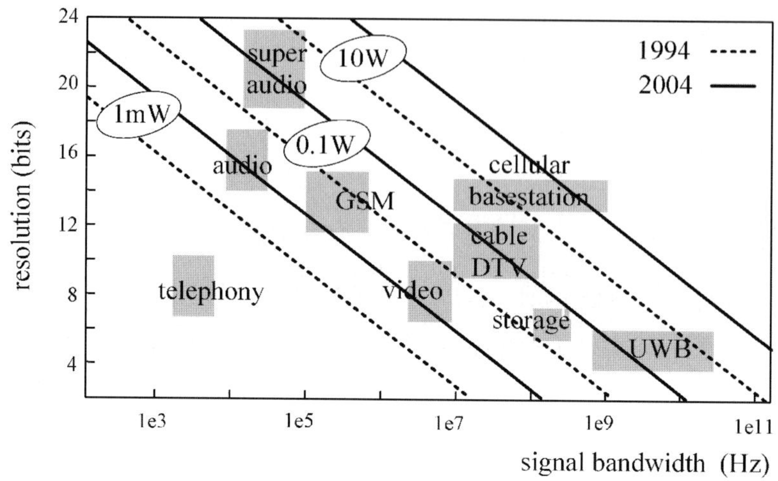

Figure 2.5: *Evolution of power consumption of ADCs for various systems [14]*

and many more.

Not only are system standards becoming more demanding, also, more and more different systems arise. These need to co-exist or even co-operate; e.g. the fourth generation of mobile communication systems targets seamless hand-over and multi-standard operation. The radios then should enable a wide range of air interfaces and the baseband conditioning channel should become very flexible as well.

2.4.3 Advances in digital signal processing and analog circuit design

Digital signal processing follows Moore's Law. This does not only translate into miniaturization, it also results in a considerable power saving. It is calculated in chapter 4, page 51 that the dynamic power consumption per transition reduces by *a factor 1000 in 10 years time.*

On the contrary, the improvement of the power/performance balance of analog circuits is much slower. In fig. 2.5, lines of constant power consumption have been added to the graph of fig. 2.4. They indicate the power consumption in the ADC for a combined resolution and bandwidth target. According to the International Technology Roadmap for Semiconductors (ITRS), these lines have shifted up by *a factor 10 every decade* [14] over the past.

Obviously, a huge gap exists between the pace of improvement in the power/performance balance of analog circuits (including data converters) and that of digital processing.

2.4.4 Digitization of the architecture

For cost and flexibility reasons, the conditioning channel is being digitized. The cost reduction aspect follows from Moore's Law; i.e.digital blocks are subject to miniaturization. The flexibility results from the fact that digital processing can easily be adapted to

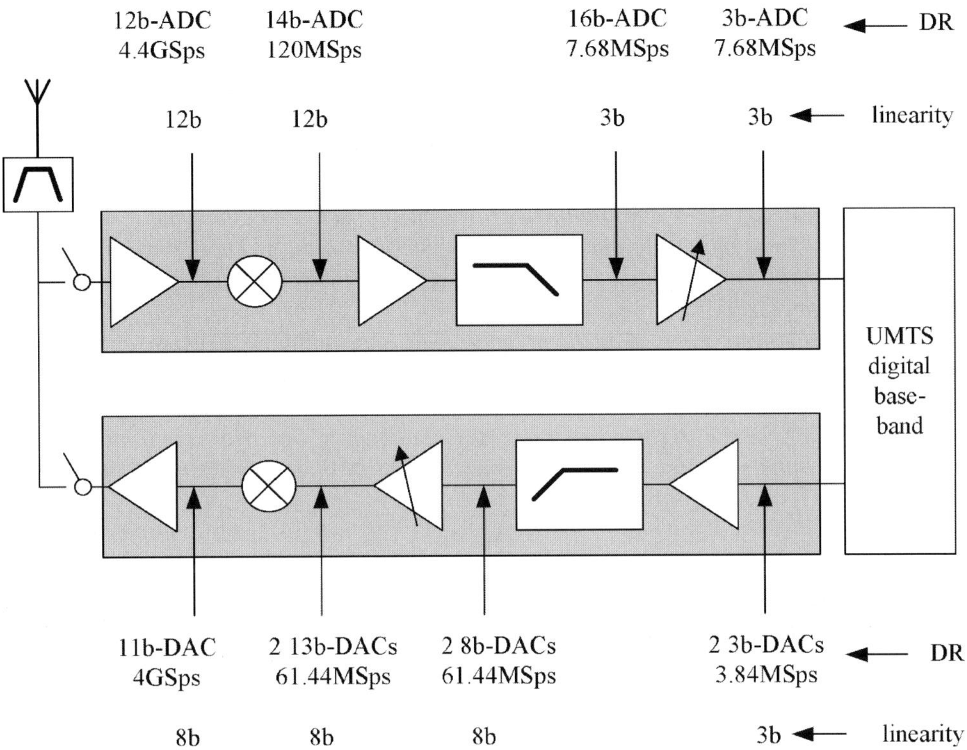

Figure 2.6: *Linearity and DR requirements on ADC and DAC with increased digitization of UMTS conditioning channels (functional diagram).*

new requirements, to encompass more features, to provide multi-mode and multi-standard operation, etc.

As such, analog conditioning circuits are being replaced by digital processing. However, while reducing the number of analog blocks, the data converter shifts towards the antenna and becomes dramatically more difficult. This is illustrated with an example next. The example treats the digitization of the A/D and D/A conditioning channels (i.e. receiver and transmitter respectively) for UMTS. First, the *DR* and *BW* requirements are derived for various positions of the ADC or DAC in the channel (corresponding to various degrees of digitization). Next, the linearity requirements are added. The calculated figures should be considered as first-order estimates only. Their prime aim is to illustrate the consequences of digitization on the converter specifications.

UMTS Example [15], [16], [17], [18], [19]:
UMTS is a spread-spectrum system with 5MHz-wide channels and a "chip rate" of 3.84MHz. The receive band from 2.11 to 2.17GHz is considered. Assuming a low IF receiver, a maximum of analog filtering and VGA, then, a 3-bit, 7.68MSps ADC suffices considering the spreading gain and the demodulation requirements. This is true both for

noise and for distortion.

Leaving out all analog VGA requires the ADC to accommodate an additional signal range of 78dB, leading to a 16-bit ADC. Since only the wanted channel is present, the linearity can remain as low as before.

If the analog filter is removed, the sample rate must be increased to prevent aliasing of interferers falling within the 60MHz-wide receive band. The resolution can be lowered by 2 bits because of the over-sampling. However, the ADC must become dramatically more linear because of the presence of in-band blocker channels. These can be as strong as −25dBm, while the distortion must remain below −103dBm. As such a linearity of 12 bits becomes necessary.

If the ADC is put at RF, near the antenna, then, the sample rate increases further to about 4.4GSps. The resolution of the ADC can be lowered to 12 bits because of the very high over-sampling.

Notice that, in the digitized receiver, the challenging linearity requirement is especially due to the presence of the interferers instead of directly relating to the wanted signal.

For the transmitter, a zero IF architecture is suitable. Essentially, a pair of 3-bit quadrature DACs sampled at Nyquist rate can be used to digitize the OCQPSK modulated signal. However, this would require infinitely steep analog filters to suppress the higher-order replicas of the baseband signal.

In practice, over-sampled DACs with higher resolution are used to meet the spectrum and the spurious emission requirements. All analog filtering (except the antenna filter) can be omitted at the expense of 16-times over-sampled, 8-bit DACs. These DACs generate quantization noise substantially below the spectrum emission limit of $-49dB_c/1MHz$ *at 12.5MHz offset. In case of straightforward multi-bit quantization, the distortion of a single 8-bit DAC is at best* $-50dB_c$*. For other encoding schemes, the quantization related distortion can be lower. Therefore, when assuming a reasonably linear implementation, the spectrum emission limit is respected with a margin in the quadrature configuration.*

The analog VGA can be traded for digital VGA at the expense of 13-bit DACs. These provide the 71dB of power control on top of the 3-bit resolution reserved for the synthesis of the modulated signal while taking advantage of the 16-times over-sampling. (At low output levels, the emission specification is set to an absolute limit and these DACs inherently meet that specification.)

Direct digital synthesis -leaving out the analog mixer too- would require a sample rate of around 4GSps and a single 11-bit DAC. (Because of the oversampling, the DAC can have about 3 bits less than before).

Notice that the linearity requirement is set by the spurious emission limit. As such, it equals the DR requirement.

Clearly, conventional, straightforward digitization, by pushing the ADC and DAC towards the antenna, puts an enormous burden on the converters[2]. In addition, new systems

[2]In fact, especially for the ADC the requirements may become even stricter than what is suggested in the example. The lack of analog gain means that the specified dynamic range must be achieved at a smaller input

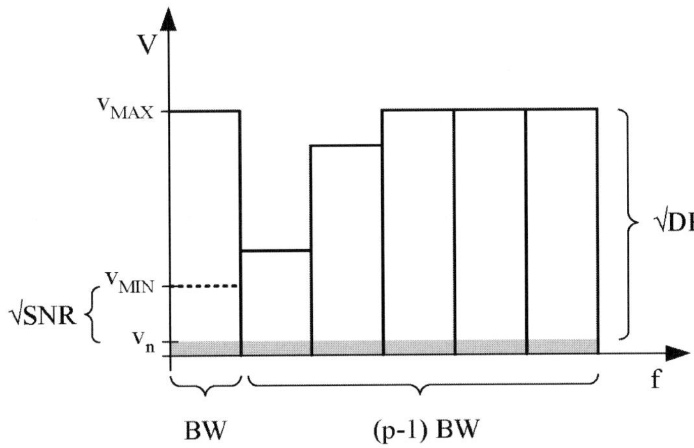

Figure 2.7: *A general input spectrum and definitions*

demand a higher performance conditioning channel and technology evolves in an unfavorable direction. The power/performance balance of ADCs improves too slowly to meet all the demands. It is recognized that *innovative channel architectures are needed to bridge the performance gap.* For instance, [20] proposes "digitally assisted analog circuit design"; i.e. requirements on analog precision are delegated to a digital processor. Alternatively, here, *the focus of the architectural innovation is on optimally exploiting the characteristics of the ΣΔ ADC and on "not wasting power to the interferers".* This is clarified throughout the book.

2.5 Nomenclature

Fig. 2.7 depicts an example input spectrum to an A/D conditioning channel. This spectrum corresponds to a typical interference scenario in communication systems. In accordance with the nomenclature in that field the notion of "wanted channel" and "interferer channels" is used. Together, these occupy the entire "signal band". In this book all metrics like BW, $\hat{v}_{\mathrm{IN}}$, $\hat{v}_{\mathrm{MIN}}$, $\hat{v}_{\mathrm{MAX}}$, $\overline{v^2}_{n,eq}$, *DR*, *SNR*, etc., refer to the wanted channel only, unless stated otherwise[3]. The symbols indicating a ratio, are sometimes expressed in decibels. Then, this is mentioned as a subscript to the symbol or it becomes clear from their value.

The following metrics are indicated in fig. 2.7.

- *BW:* this is the bandwidth of the wanted channel. The bandwidth of the entire band is larger by a factor p ($p \geq 1$). By consequence, a maximum of $p-1$ interferer channels can be present in the band.

signal than in the conventional channel.

[3]For simplicity reasons, the signal at any node in the channel, is often characterized by its amplitude only. In those cases a high impedance is assumed or only a ratio of amplitudes is important. If this is not the case, the corresponding impedance value should be mentioned as well.

- $\hat{v}_{\mathrm{IN}}$, $\hat{v}_{\mathrm{MIN}}$ *and* $\hat{v}_{\mathrm{MAX}}$*:* these correspond to the instantaneous, the minimum and the maximum amplitude of the wanted signal respectively. The amplitude levels $\hat{v}_{\mathrm{MIN}}$ and $\hat{v}_{\mathrm{MAX}}$ are introduced to allow a discussion on the consequences of variable gain implementation. It should be stressed that all refer to the amplitude of the signal in the wanted channel only. In case interferer signals are present, the total signal amplitude will be larger by a factor q ($q \geq 1$) because the interferers and the wanted signal cannot easily be distinguished from each other in the time domain. The overall amplitude then equals $q\,\hat{v}_{\mathrm{IN}}$.

- v_n*:* this symbol is used as a short-hand notation for the rms value of an input-referred, equivalent noise source corresponding to the wanted channel only.

- *DR:* the dynamic range is the ratio of the maximum signal power to the minimum detectable signal power in the channel bandwidth. The minimum detectable power -theoretically- approaches the minimum noise level. Then, the dynamic range also equals the ratio of the maximum signal power to the integrated idle channel noise.

- *SNR:* the signal/noise ratio is the ratio of the power of the wanted signal to the noise power[4] is applied. Both are integrated over the channel bandwidth and both are present simultaneously. This specification is set by the digital processing. For example it can originate from a requirement for digital demodulation with a certain bit-error rate.

In this book, it is understood that the above explained metrics refer to the wanted channel only (unless stated otherwise).

Fig. 2.7 does not show the linearity specification. In this book, we consider intermodulation distortion instead of harmonic distortion (both are related and can be calculated from each other). Harmonic distortion results in components at higher frequencies, likely outside of the bandwidth of the wanted channel. Hence, these can be suppressed by filtering. On the contrary, intermodulation distortion may cause components inside the bandwidth of the wanted channel. During consecutive conditioning or processing, the distortion components cannot be discriminated anymore from the wanted signal. As an additional argument, in many system standards, intermodulation tests are defined explicitly. As such, this is a meaningful distortion parameter for the conditioning channel as well.

Assuming differential operation, we use IM_3 to quantify the distortion. The IM_3-definition is illustrated in fig. 2.8:

- IM_3*:* is defined for a two-tone test in which two sine wave inputs, of equal amplitude and at nearby frequencies f_1 and f_2 are applied. The IM_3-value equals the ratio of the amplitude of the intermodulation component at $2f_2-f_1$ (or at $2f_1-f_2$) to that of the fundamental signal at f_1 (or at f_2). In fact, this value depends on the signal level. In case this signal level is not mentioned it is understood that the fundamental signals both have an amplitude of half the allowed full-scale amplitude (thus at $-6dB_{FS}$).

[4] For some systems, e.g. using time-division multiplexing, this definition can be adapted in an obvious manner.

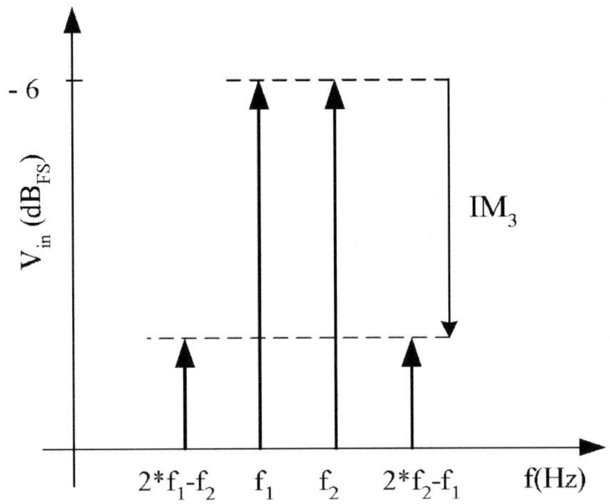

Figure 2.8: *Definition of IM_3*

In contrast with the above parameters, *the IM_3 distortion of the various blocks is not solely linked to the wanted signal. It depends on the overall signal amplitude consisting of both wanted and interferer signals. Often, the interferer signals cause the dominant distortion.*

In many systems (like the example of digital demodulation) distortion and noise -in a first order approximation- have a similar effect on the processing and both must be equally low. Then, the following relation between the IM_3-specification and the DR-specification can be derived:

$$IM_3 \Leftrightarrow \frac{\overline{v}_n\sqrt{2}}{\hat{v}_{\mathrm{IN}}/2} = \frac{2}{\sqrt{DR}} \tag{2.2}$$

In case values are filled out, the symbol $\Leftrightarrow$ can be replaced by $=$. *Also note that DR is used and not SNR.* This follows from the assumption that the major distortion problem is due to strong interferers in the presence of weak wanted channels.

Often, the distinct contribution of noise and distortion is not important and the signal-to-noise-and-distortion ratio (*SINAD*) is used. In those cases, *SINAD is used when assuming noise and distortion contribute equally.*

2.6 Conclusions

From Shannon's theorem, and considering practical constraints on analog circuit design, noise, distortion and bandwidth are identified as key parameters for evaluating the channel performance for a given input signal.

Due to demanding new systems and the evolution of CMOS technology, the implementation of the conditioning channel becomes very challenging.

Digitization of the conditioning channel, further challenges the performance of the ADC.

From the past, the power/performance balance in analog circuits and ADCs improves by a factor of 10 per decade. In the same period of time, the power/performance balance in digital processing, improves by a factor of 1000.

Architectural innovation is identified as a degree-of-freedom to meet the requirements on the conditioning channel.

In highly-digitized conditioning channels for wireless communication systems, the ADC specifications become increasingly determined by the presence of the interferers instead of only relating to the wanted channel.

Chapter 3

ΣΔ A/D conversion

This chapter starts with a historical overview of ΣΔ A/D conversion. Next, the state-of-the-art in ΣΔ converter design is briefly discussed. This discussion includes some architectural choices and circuit aspects of a ΣΔ design. It provides a common ground for use in later chapters. In addition, it is motivated that a single-bit, continuous-time implementation of a ΣΔ ADC with a feed forward type of loop filter, is an inherently low-power topology.

Considering the discussion in chapter 2, this type of ΣΔ ADC has the prospect of becoming an enabling block in future conditioning channels. This observation is a basis for most of the book.

Finally, for completeness, some limitations on the use of ΣΔ ADCs are summarized.

3.1 Historical overview

A first patent related to delta modulation was filed in 1948 [21]. In 1952, the delta modulator was first published by de Jager and Greefkes at the Philips Research Laboratories (fig. 3.1, [22]). The invention was inspired by the operation of the human brain: physiological signals are translated into a series of electrical pulses in the nerve system as a means for data transmission to the brain. In a similar way, pulse density modulation was used for robust data transmission in telephony. In the single-bit, delta-modulated code all bits are of equal weight and any bit-flip causes only a small error. In a multi-level PCM code though a bit-flip of the Most-Significant-Bit results in a major error. The paper by de Jager in 1952 [22], on delta modulators was the first in a massive series of delta-sigma and sigma-delta papers. In 1960, the delta-sigma modulator was patented by Cutler [23]. Inose et al. [24] proposed to shift the loop filter in the forward path of the modulator in 1962. In the 1980's sigma-delta conversion became popular in both the A/D and the D/A part of audio channels. In addition, instrumentation applications widely adopted ΣΔ converters. By that time, a lot of theoretical work had been published, a.o by J. Candy.

In 1993, for the first time a separate session at the ISSCC was devoted to ΣΔ A/D conversion. The target applications soon evolved including digital radio and 2G com-

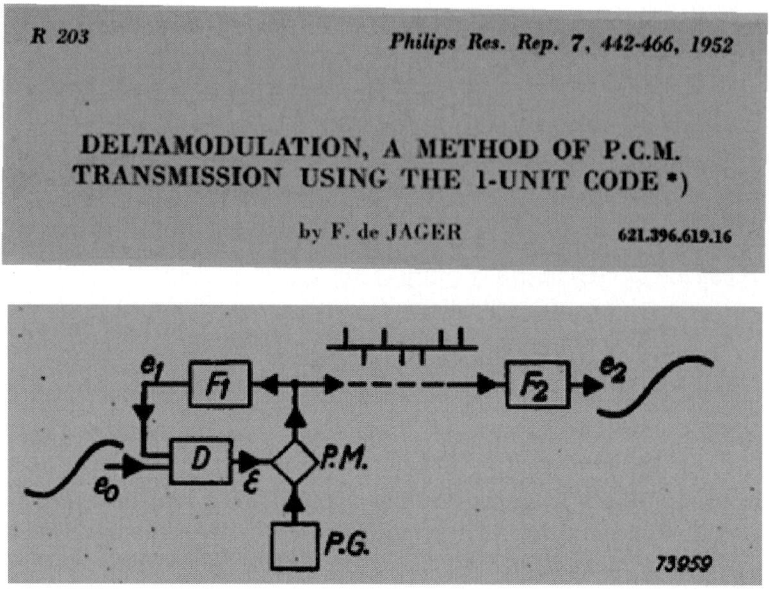

Figure 3.1: *The delta modulator as first published by de Jager and Greefkes in 1952*

munication standards. From 2000 on, designs for 3G communications and for wireless connectivity in the 2.5GHz band have been widely published. Anno 2005, also wireless connectivity in the 5GHz band and advanced wired communication standards -like ADSL- are targeted.

3.2 State-of-the-art in ΣΔ A/D conversion

This section gives an overview of some architectural and implementation related aspects of the ΣΔ design. As such, it provides a common ground for later reference. For a thorough analysis of ΣΔ data converters the reader is referred to a.o. [25], [26], etc.

A formula for the signal/quantization-noise ratio (*SQNR*) of a ΣΔ modulator is repeated here because it gives an overview of the various architectural parameters and their effect on *SQNR*:

$$SQNR = \frac{3}{2}\left(\frac{2L+1}{\pi^{2L}}\right)\left(2^N - 1\right)^2 m^{2L+1} \tag{3.1}$$

where L = order of the loop filter
N = number of bits in the quantizer
m = the over-sampling factor; i.e. $m = \frac{\text{actual sample rate}}{\text{Nyquist sample rate } f_s}$

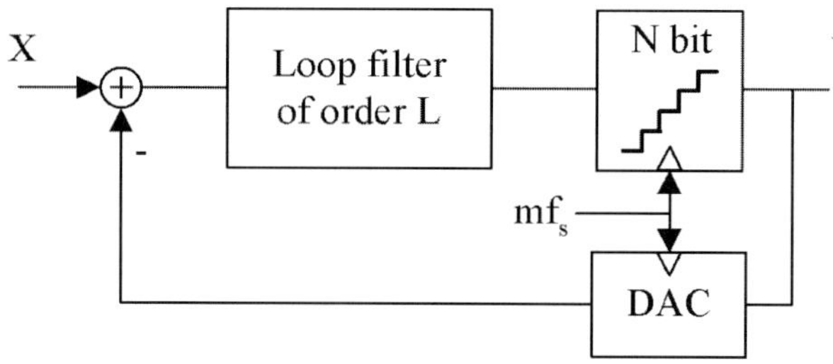

Figure 3.2: *Block diagram of a ΣΔ ADC and architectural parameters*

These parameters are further clarified in the block diagram of fig. 3.2. They provide the framework for the discussion on the architectural design in the next section.

It should be remarked that eq. 3.1 results from theoretical calculations. It does not take into account stability issues. For this reason, the achievable *SQNR* may, in practice, be lower than what is predicted from eq. 3.1.

3.2.1 Architectural considerations

Normally, a ΣΔ ADC is dimensioned such that its quantization noise (within the targeted bandwidth) is lower than its circuit noise. The circuit noise provides dithering and reduces the appearance of tones. In addition, this dimensioning also yields lower overall power consumption for the ADC. While a reduction of the noise power of an analog circuit requires a linear increase in power consumption, the quantization noise can be lowered by more power-effective, architectural measures. These architectural degrees-of-freedom are discussed next.

Over-sampling factor

In an industrial realization, the value of the over-sampling factor is often dictated by the system. For instance, the sample rate may be standardized or a specific clock frequency is available and should be re-used. (On the contrary, the order of the loop filter and the choice of a filter topology, as well as the number of bits in the quantizer are mostly degrees-of-freedom in the design of the ΣΔ ADC.) The choice of the over-sampling factor has a strong impact on the *SQNR*. Next to appearing explicitly in eq. 3.1, also L is a function of m. This is discussed below.

Loop filter

The architectural design of the loop filter includes the choice of the filter order, the choice between a switched capacitor or a continuous-time implementation and the choice of filter topology.

Order of the loop filter: For a fixed over-sampling factor, the highest sensible filter order should be implemented, because:

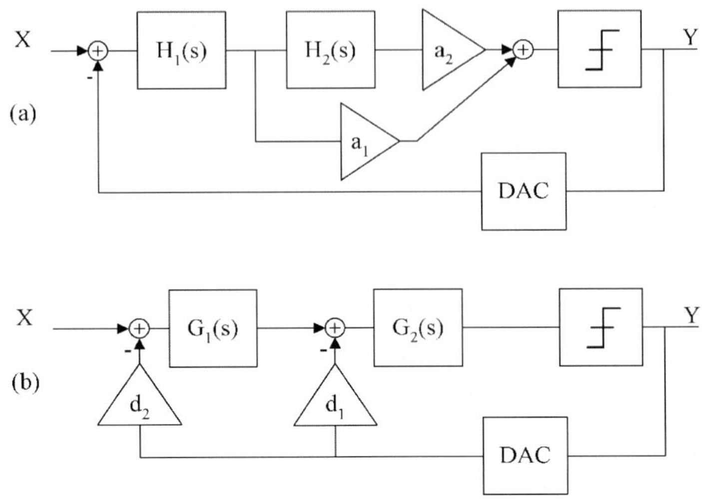

Figure 3.3: *Continuous-time ΣΔ ADC with feed forward (a) and feedback (b) loop filter*

- the higher-order filter sections provide additional shaping of the quantization noise;
- still, the power consumption of these sections can be very low: since the filter is put inside of an overall feedback loop, the noise and distortion of these stages is suppressed by the gain of the preceding sections.

On the other hand, stability issues put an upper boundary on the filter order.

Switched capacitor versus continuous-time implementation: Key differences between switched capacitor and continuous-time implementations are discussed in [27]. A switched capacitor implementation is more robust to clock jitter and processing spread and its coefficients can be derived from those of a digital modulator in a straightforward way. A continuous-time implementation often performs better with respect to power consumption. In addition, it features implicit anti-aliasing filtering.

Feedback or feed forward loop filter: Fig. 3.3.a and 3.3.b respectively depict a ΣΔ ADC with feedback and with feed forward compensation (for high-frequency stability of the loop). Both types of loop filter provide the same noise-shaping but a different closed-loop transfer. (In fact, a mixture of feed forward or feedback paths can be implemented as well.) In case of feedback compensation all filter stages need to linearly integrate the entire output signal of the DAC. In case of feed forward compensation, the input signal to the higher order filter stages has been filtered by the preceding stages relaxing the linearity and bandwidth requirements. Therefore, the feed forward implementation consumes less power. These differences are further discussed in 6.3. As a short-hand notation these topologies are referred to as "feed forward ADC" and "feedback ADC".

Number-of-bits in the quantizer

Multi-bit quantization provides a means for increasing resolution without increasing the filter order or sample rate. This translates in a higher maximum stable-input range, lower quantization noise and a better defined quantizer gain than for a single-bit design. Hence, stability is achieved more easily in multi-bit modulators.

Single-bit quantization has the advantage of inherent linearity while conventional multi-bit designs need "dynamic element matching"(DEM) of the DAC. The implementation of the DEM-algorithm, of the multiple comparators and of the DAC represents a complexity -almost- quadratic with the number of bits. Hence, the area of a multi-bit design is larger in those cases. Recently, alternative multi-bit designs have been presented: [28] uses a multi-bit semi-digital filtering DAC to alleviate the linearity problem, in [29] the output of the DAC is truncated in order to avoid a DEM algorithm.

The comparison of the power consumption is somewhat less straightforward: circuit noise requirements are the same for the multi-bit and single-bit design, the hardware complexity of the multi-bit design is largest but the active circuits in the single-bit design need a higher bandwidth (because of the higher sample rate). Fortunately, the bandwidth increase in the single-bit case is moderate: for a same *SQNR* a single-bit design needs only a slightly higher sample rate than a multi-bit design (eq. 3.1). This implies a proportional -thus moderate- increase in bandwidth and current consumption for the active stages. Balancing this with the complexity in the multi-bit design single-bit sigma-delta ADCs are likely to achieve a better power efficiency for most applications. In fact, this is in line with the discussion on the Shannon theorem and power efficiency of analog circuits (page 29): resolution in amplitude (e.g. multi-bit quantization) leads to a higher power consumption than resolution in time (e.g. single-bit quantization at large over-sampling). Especially the DAC of the multi-bit design constitutes a significant part of the power consumption because noise and distortion of the DAC are directly present at the output of the ΣΔ ADC.

In a summary, multi-bit modulators are easier to design in view of stability while single-bit designs outperform in terms of power/performance for most applications. In section 3.4, it is discussed how the performance of single-bit ΣΔ conversion degrades at high sample rates due to various non-linear effects (jitter, inter-symbol-interference, etc). At these sample rates (typically required for applications with a large signal bandwidth and a high *DR* specification) multi-bit designs become more attractive. At an equal target performance, multi-bit modulators can be sampled at a -moderately- lower frequency (see above). In addition, multi-bit quantization reduces the influence of the mentioned effects on the overall ADC performance.

Examples
Table 3.1 compares power and performance of a few sigma-delta designs for single-bit and multi-bit implementations[1]*. All achieve high resolution. The power efficiency of the*

[1] Only designs that are among the best published have been included in this comparison. In addition, the pairwise comparison is between designs in the same technology generation.

Table 3.1: *Comparison of single-bit and multi-bit ΣΔ ADCs for a few conversion bandwidths*

Conversion bandwidth of ~ *200kHz*

Design	*DR*	*SINAD*	*P*	*FOM* (eq. 4.4)
1-bit: [30]	92dB	90dB	8mW	3×10^{-17}J
4-bit: [31]	75dB	64dB	1.75mW	3×10^{-15}J

Conversion bandwidth of *1MHz*

1-bit: [32]	76dB	75.5dB	4.4mW	1×10^{-16}J
4-bit: [33]	68dB	64dB	2.2mW	1×10^{-15}J

Conversion bandwidth of *15MHz*

4-bit:[34]	67.5dB	63.7dB	70mW	2×10^{-15}J

designs is compared using a figure-of-merit (FOM) that is discussed in section 4.2. This FOM is the ratio of the power consumption of the design to the product of SINAD and BW. Hence, the lower the FOM the better the design. For a conversion bandwidth lower than about 2MHz single-bit sigma-delta conversion achieves a better power/performance ratio. For conversion bandwidths of 15MHz or higher only multi-bit designs (often in combination with a MASH architecture) have been published.

Summarizing the above, ΣΔ ADCs with a continuous-time loop filter built on a feed forward topology and using single-bit quantization, constitute an inherently low-power architecture.

3.2.2 Implementation aspects

For later reference, some considerations on the requirements on sub-blocks of the ΣΔ ADC, and some possible circuit topologies are briefly discussed here. The block diagram of fig. 3.4 shows an example implementation of a single bit, continuous-time ΣΔ ADC with second-order, feed forward loop filter. This block schematic is a popular starting point for many continuous-time ΣΔ ADCs (a.o. [7], [35], [36], [37], [38] and [39]) as well as for the designs presented in this book.

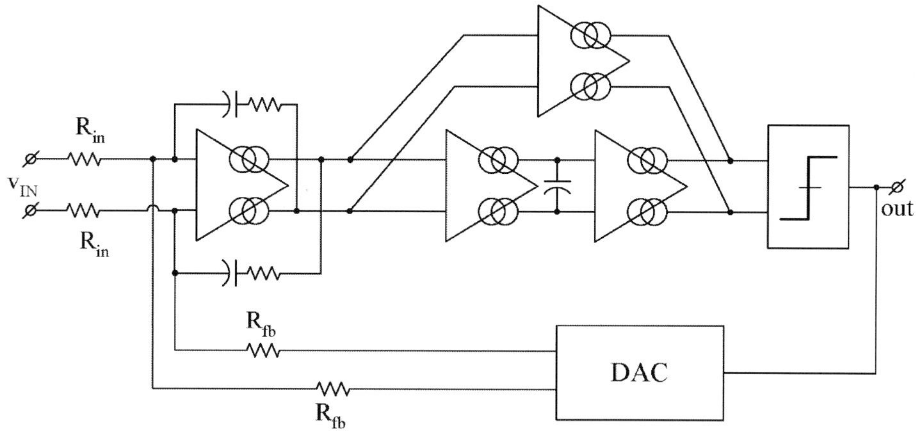

Figure 3.4: *Continuous-time ΣΔ ADC with feed forward loop filter*

V/I conversion in the input and in the feedback path

In fig. 3.4 resistors perform the linear V/I conversion in the input and in the feedback path. In [7] operational transconductance amplifiers (OTAs) are used. These can achieve a lower phase-shift compared to the resistors that act as a lumped RC-network. The OTAs are less linear though.

Loop filter

In case of resistive V/I conversion the input and the feedback current are summed at the virtual ground node provided by an OTA in an integrating, negative feedback configuration. This OTA constitutes the first integrator. In case OTAs are used for the V/I conversion the currents can be summed on the first capacitor. For both implementations the major requirements on the first stage relate to noise and distortion The second integrator -and all following integrators in case of a higher-order filter- could be based on a similar stage or alternatively it consists of a g_mC-section (as depicted). In the latter case, degeneration may be applied for linearity reasons. Parasitic poles of the integrators should be at high frequencies in order not to add phase delay to the loop. Therefore, OTAs are widely used. In a mixed-signal environment the OTA is normally based on a differential pair. In addition, the input transistors can be cascoded for linearity reasons. Possible topologies and their specific advantages are:

- OTA with telescopic cascode: requires only a minimum number of current branches;
- OTA with folded cascode: allows a large output swing for the integration.

In the latter topology the power consumption is doubled because of the double amount of current branches. Therefore, the designs presented in this book use a telescopic cascode topology. Still, if the supply voltage of future CMOS technologies continues dropping it may become necessary to use a folded cascode topology instead.

Other topologies based on a differential pair include:

- differential gain stage followed by source-follower: provides a low output impedance for fast settling;
- a cascade of two differential gain stages with Miller compensation: provides high DC-gain.

These topologies have an additional bandwidth limitation on the internal node between the consecutive stages. Therefore, they are not preferred for ΣΔ A/D conversion at a high sample rate.

Quantizer and DAC

The quantizer is not critical due to the loop operation. It should not add significant delay though. The feedback DAC consists of switches connecting the resistors to the negative or the positive reference voltage depending on the polarity of the output data.

Further detail on possible implementations can be found in a.o. [40]. Notice that the proposed implementation allows technology scaling and operation at a low supply voltage: only the input stage is critical. It needs to have a reasonable output swing without going into saturation. Fortunately, its input is a virtual ground node and therefore the input signal swing is negligible.

3.2.3 Performance metrics for ΣΔ ADCs

The performance of a ΣΔ ADC with respect to noise and stability is typically characterized in a diagram as depicted in fig. 3.5. The *SNR* increases with the input level until the peak-*SNR* is reached. Beyond this input level the quantization noise rises and the *SNR* degrades because of a reduced stability. The mechanism behind this large-signal instability is the fact that the DAC cannot feed back a large enough signal in time to compensate for the input signal (i.e. the phase and the gain margin of the loop become too small). The error signal and the internal signals in the loop (i.e. the outputs of the various integrators) therefore grow and further reduce the gain and the phase margin resulting in instability. For a further discussion on stability the reader is referred to [25], [40].

In this book, the input level corresponding to the peak-*SNR* is defined as the maximum stable input. The *DR* is defined relative to this input level. The designs presented in this book (chapter 8) typically achieve a maximum stable input corresponding to a modulation depth of the digital output of $\sim 70\%$[2]. This is *3dB* below digital full-scale. This modulation depth corresponds to a good compromise between aggressive noise shaping and stability. Designs achieving a higher modulation depth have a lower peak-*SNR*. Vice versa, designs with a higher peak-*SNR* only allow a smaller modulation depth[3].

[2] The electrical value of the maximum stable input is implementation specific: next to the allowed digital modulation depth it also depends on the output current/voltage of the DAC in the feedback path.

[3] Notice, in some texts the definition of *DR* is based on a -theoretical- 100% modulation depth compared to the idling noise. Especially for designs with a poor stability this definition results in misleading, high values for both *DR* and peak-*SNR*.

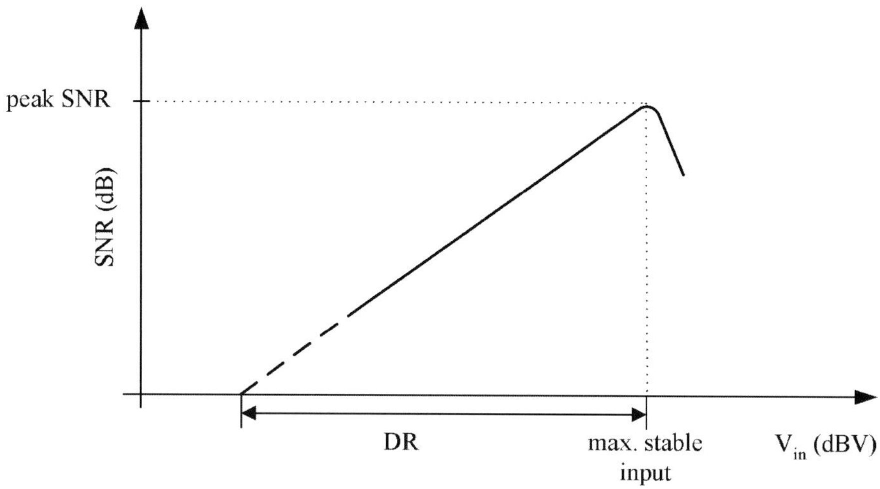

Figure 3.5: *Definition of some performance metrics for ΣΔ ADCs*

3.3 ΣΔ ADCs in future conditioning channels

Considering the discussion in chapter 2 on the playing field for future conditioning channels, it is motivated that ΣΔ ADCs can be a key enabler for these channels. It is shown that ΣΔ ADCs are built on an inherently low-power architecture and that this is due to two main characteristics; i.e. *over-sampling is used* and *overall feedback is applied.* The first bullet relates to the fact that ΣΔ modulators trade resolution in amplitude, i.e. *SINAD*, for resolution in time, i.e. *BW*. It is mentioned in section 2.4.1, that for this reason, ΣΔ based conditioning channels seem attractive for integration in advanced CMOS technologies. Here, it is added that, in view of Shannon's theorem and in view of power consumption in analog circuits, this is an asset leading to a low-power solution.

The second bullet is important for a comparison with Nyquist A/D converters. The overall feedback relaxes various accuracy requirements on analog sub-blocks.

Finally, the previous, rather intuitive reasonings are supported by a survey of published power/performance of A/D converters.

3.3.1 The Shannon theorem and ΣΔ based signal conditioning

The Shannon theorem considers the data rate through a channel. Here, it is applied in a more generalized way; i.e. on an electrical signal. This signal -just as the Shannon channel- is characterized by a bandwidth and a *SNR*. These parameters determine the data rate that the signal can contain. For sure, the signal properties may not fully meet the Shannon assumptions (a.o. Gaussian signal and noise sources), but this discussion is meant as an intuitive reasoning only.

Hence, applying Shannon's theorem to a true-life signal:

$$\text{data rate} \sim BW$$
$$\text{data rate} \sim \log SNR$$

where data rate, *BW* and *SNR* relate to the signal. This means that, for a fixed data rate, resolution in amplitude, i.e. *SNR*, can be exchanged for resolution in time, i.e. bandwidth. This property is the basis of ΣΔ encoding.

Taking into account the following considerations on power consumption in analog circuits:

$$\text{power consumption} \sim BW$$
$$\text{power consumption} \sim SNR$$

assuming the latter is dominated by thermal noise. Here, power consumption, *BW* and *SNR* relate to the circuit.

The bandwidth and *SNR* requirements on the analog circuit are determined by the bandwidth and the *SNR* of the signal. This links the power consumption of the analog circuits in the conditioning channel to the ΣΔ encoding of signals. Hence, it can be concluded that:

- in view of power-efficiency of the analog conditioning circuits, an encoding scheme that trades resolution in time for resolution in amplitude, should be used.

This is exactly what happens in a ΣΔ modulator.

3.3.2 Comparison of Nyquist and ΣΔ based signal conditioning

A conditioning channel consisting of an anti-alias filter and a Nyquist A/D converter is depicted in fig. 3.6.a. In fig. 3.6.b, the block diagram of a conditioning channel with ΣΔ A/D conversion is shown. Because of the over-sampling, the ΣΔ ADC reduces the requirements on the anti-alias filter. In fact, it is analyzed in chapter 5, and demonstrated with examples in chapters 8 and 9, that in many ΣΔ based conditioning channels the anti-alias filter can be omitted. As such, it is not included in the present analysis either.

The loop filter and the quantizer of the ΣΔ ADC, can be compared to the anti-alias filter and to the Nyquist ADC, respectively.

The specifications on the anti-alias filter in the Nyquist channel are challenging:

- all stages of the filter -or generalized, all stages of the Nyquist channel- need to achieve the full dynamic range requirement[4]. Even though the consecutive stages can have a higher noise (inversely proportional to the preceding gain), they must remain linear over a larger input range, as well;

[4]For simplicity reasons, variable gain amplification is not considered in this comparison, even though, it would reduce the dynamic range requirements for following blocks. It is shown in chapter 6 and 9 that variable gain can also be implemented in ΣΔ ADCs, yielding a similar benefit.

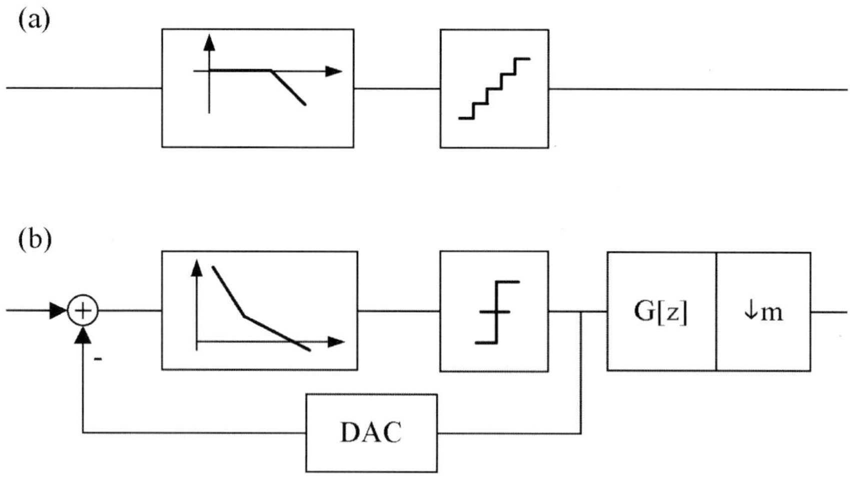

Figure 3.6: *Comparison of an analog conditioning channel with Nyquist ADC (a) to a single-bit, ΣΔ based conditioning channel (b)*

- the accuracy requirements on the anti-alias filter in the Nyquist channel can be very stringent. These requirements relate to the pass-band ripple, the transition band and the suppression required at half the sample rate.

On the contrary, the design of the loop filter for the ΣΔ ADC is much more relaxed:

- the consecutive filter stages of the ΣΔ loop filter can have increasing noise and distortion. Because of the overall feedback, these contributions are suppressed towards the output;

- the accuracy of the loop filter transfer is not too critical. The gain of the loop filter needs to be "high"within the conversion bandwidth, while the accuracy of the gain is not important. A deviation on the cut-off frequencies can be taken into account when defining the gain and phase margin of the loop.

For completeness, it is mentioned that the DAC in the feedback path of the ΣΔ modulator is a critical block. Still, for a single-bit design the required accuracy is normally easily achieved. On the contrary, the requirements on the Nyquist ADC are strict:

- the number-of-bits and the accuracy of the sub-blocks of the ADC directly affect the resolution in the channel;

- theoretically, the Nyquist ADC can be sampled at a frequency equaling twice the signal bandwidth; i.e. $f_s = 2BW$. In practice, $f_s > 2BW$ is required in order to relax the transition-band for the anti-alias filter and in order to reduce the amplitude and phase error of the sampling action.

The quantizer in the ΣΔ ADC can be inaccurate:

- the number-of-bits in the quantizer can be lowered (ultimately to a single bit) because the quantization noise added in the quantizer is shaped by the transfer of the preceding loop filter. Similarly, all other noise and distortion sources in the quantizer are not critical either;
- the sample rate is higher, but since CMOS-technology is optimized for high-speed switching, this should not be a problem for most applications (see discussion in section 3.4.

In order to make a completely fair comparison, the power consumption of the decimation filter in fig. 3.6.b should be taken into account as well (see section 4.5). The above comparison, does indicate that the use of ΣΔ A/D conversion significantly relaxes the accuracy requirements for most analog blocks in the channel. This is confirmed by the comparison of published power/performance for both types of A/D converters, below.

3.3.3 Survey of published power/performance values

The power/performance balance of various converters can be benchmarked using the Figure-of-Merit (FOM) according to eq. 4.6: the lower the FOM-value, the better the ADC. This FOM is defined and motivated in section 4.2. Here, it is used for a first-order comparison of the power/performance balance of ADCs that may differ both in architecture, implementation style, technology, etc.

The data for this analysis is taken from [41], evaluating this FOM for Nyquist ADCs, and from table A.1 in appendix A, listing this FOM for discrete-time and continuous-time ΣΔ ADCs. From this data, the following is observed:

- for Nyquist ADCs, the best FOMs equal $\sim$ 1pJ or higher;
- for discrete-time ΣΔ ADCs, FOMs of 0.35pJ and 0.5pJ have been reported ([42] and [43] resp.), but most exceed 0.9pJ;
- for continuous-time ΣΔ ADCs, a lot of designs achieve a FOM in the range of 0.3pJ to 0.6pJ .

This comparison favors continuous-time ΣΔ A/D conversion[5] and confirms the statements of sections 3.3.1 and 3.3.2. In fact, the lowest FOM is achieved for single-bit feed forward implementations. This was anticipated in section 3.2.1.

3.4 Limitations of ΣΔ A/D conversion

From the above, ΣΔ ADCs seem to enable low-power, high performance conditioning channels that are scalable to deep-submicron CMOS technologies. Historically, the ap-

[5] For a fair comparison, the power consumption in the decimation filter of the ΣΔ designs should be taken into account as well. However, technology scaling is very helpful in reducing this contribution.

plication area of ΣΔ converters has been restricted to low bandwidth, high resolution systems. Even in advanced technologies, though, the achievable sample rate is hampering the adoption of ΣΔ conversion for high-performance, wide-band applications. In the discussion, linear and non-linear limitations are distinguished in the sense that linear limitations (such as phase margin) do scale with technology and current consumption, while non-linear limitations must be overcome by a more than proportional power increase or by innovative circuit design.

3.4.1 Linear limitations

The sample clock in a ΣΔ ADC -by definition- is significantly higher than the signal bandwidth. It sets the time interval in which the feedback pulse needs to be processed by the loop filter and applied to the comparator, timely for the next sample moment. Hence, the sample frequency sets the requirement on the circuit bandwidth. This is different from an entirely analog feedback loop where the bandwidth requirement on the circuits is relative to the signal bandwidth. Some important *practical* bandwidth limitations include the limited unity-gain-bandwidth of the analog circuits, delay in the feedback path (time-constants of wiring, resistors, etc.), delay in the quantizer, etc. These limitations reduce the phase margin of the loop. Consequences are a decreased stability (a lower maximum input signal) and a reduced noise-shaping (an increase of the in-band noise), resulting in a smaller dynamic range. Some of the listed limitations can be solved by increasing the quiescent current. In addition, various methods have been published to compensate for the parasitic delay. These methods use pulse positioning of the DAC or add compensating zero-ing paths in the loop filter ([44], [45]).

Ultimately, a *fundamental* limitation on the sample rate of the ΣΔ loop is due to the f_T, f_{max} or any other frequency characterizing the maximum operation of the transistors[46].

A survey of ΣΔ papers at the ISSCC conferences from 1999 until 2003 gives an impression of state-of-the-art sample-rates for ΣΔ ADCs in a few generations of CMOS technologies (see table 3.2). It also lists a brief description of the ΣΔ architecture. Apparently, sample rates are limited to a few 100MHz (in CMOS, for high performance and robust designs) and do not really increase in advanced technologies. The reason for this discrepancy should be looked for in additional, non-linear limitations.

3.4.2 Non-linear limitations

As the sample rate of the ΣΔ ADC increases the pulse duration of the feedback DAC drops and some parasitic effects become more important: jitter on the DAC clock, inter-symbol-interference in the DAC, etc. The impact of these effects is strongly linked to the actual implementation of the DAC. Especially single-bit switched current DACs using square wave output signals are prone to clock jitter. DACs with reduced jitter sensitivity are presented in [50], [51] and [30] respectively using a cosine wave and a switched capacitor discharge pulse for the DAC. Inter-symbol-interference problems can be solved

Table 3.2: *State-of-the-art sample rate for ΣΔ ADCs in various CMOS technologies*

Reference	Technology	Sample rate	Architecture
[6]	0.13μm CMOS	*160MHz*	Mash 2-2, DT[a]
[37]	0.18μm CMOS	*153MHz*	Low-pass, CT[b]
[47]	0.25μm CMOS	*185MHz*	Low-pass, DT
[48]	0.35μm CMOS	*400MHz*	Band-pass, CT + DT
[49]	0.6μm CMOS	*100MHz*	Low-pass, CT

[a] DT=Discrete Time
[b] CT=Continuous Time

by differential operation and return-to-zero schemes. [52] and [28] present DACs with increased robustness to both inter-symbol-interference and jitter.

One other sample rate related problem in a ΣΔ ADC may be due to a limited slew-rate of the integrators. This results in non-linear, incomplete settling and therefore distortion [53]. However, slewing is easily prevented in most continuous-time designs since the peak currents are moderate compared to the quiescent current. For example, in the designs presented in chapter 9, high-frequency current transients are less than $160\mu A$ (peak-to-peak and differential) for the first integrator while its quiescent current is as high as $500\mu A$[6].

As more (and different) bandwidth limitations become important the current consumption may show a more than linear growth. Innovative circuit design is needed to overcome these limitations and to further extend the application area of ΣΔ ADCs to higher sample rates and higher resolution. A more elaborate study of these limitations is outside of the present scope.

Finally, the adoption of ΣΔ A/D converters, is *also hampered by the fact that there exists no solid proof of their stability.* Similarly, some aspects of the ΣΔ converter remain difficult to analyze. Again, this is outside of the present scope.

3.5 Conclusions

Next to providing a common ground for later reference, the discussion in this chapter leads to the conclusions listed below.

[6] In switched-capacitor designs, slewing is more likely to occur due to the large charge-transfer transient currents. However, the resulting distortion can be kept low if there is sufficient time to settle to the correct final value [54].

$\Sigma\Delta$ ADCs with a continuous-time loop filter, built on a feed forward topology and using single-bit quantization, constitute an inherently low-power architecture.

The requirements for $\Sigma\Delta$ ADC circuit design, match the characteristics of advanced CMOS technologies; i.e. most circuits are nor critical with respect to distortion, noise or matching but they do need a high bandwidth.

The basic reasons for the excellent power/performance of $\Sigma\Delta$ ADCs are the use of over-sampling and the use of overall feedback.

In a survey of published ADCs, the power/performance balance of continuous-time $\Sigma\Delta$ ADCs outperforms that of other classes of ADCs.

The application area of $\Sigma\Delta$ A/D conversion is limited by sample rate related problems. At present, sample rates of a few *100MHz* are feasible in standard CMOS.

Chapter 4

Power consumption in channel building blocks

The aim of this chapter is in analyzing the power consumption of the ΣΔ based conditioning channel as a function of the performance parameters. First, a brief review of literature on this topic is presented and some commonly used "figures-of-merit" are introduced. Next, a dedicated analysis of the power consumption in the major building blocks of the channel is conducted. A power/performance relation is derived for the analog circuits, for the ΣΔ ADC and for the decimation filter separately. The comparison of the different power/performance relations yields an understanding of possible tradeoffs when exchanging analog for digital conditioning. As such, these results are used for the analysis and comparison of various channel architectures in chapters 5 and 6. The power/performance analysis of the ADC supports the use of one particular figure-of-merit -and a small correction thereon- throughout the book.

4.1 Literature on power/performance analysis

In literature, many articles on power consumption versus performance are found. Calculations on *fundamental limits* for the power consumption of *a single transistor or a basic cell* can be found in a.o. the work of Vittoz [55] (calculating the power consumption per pole as a function of dynamic range for analog) and the work of Meindl [56] (giving the power consumption for a single binary transition).

Other references present similar calculations on basic *circuits*. [57] calculates the minimum power consumption of a continuous-time ΣΔ A/D converter limited by the noise of the input and the feedback resistor. [5] analyzes the absolute minimum power consumption of analog circuits as a function of a specific *SINAD* over the full signal bandwidth. In [8], [58] the model is refined by including the circuit's intrinsic distortion due to a finite output impedance.

Many power/performance relations and *extrapolated predictions* on circuit performance have been published. [1], [4], [59], [12] and many other references treat power requirements due to limitations on matching, bandwidth, dynamic range, etc.

Most of the above referenced publications are not adequate to conduct the power/performance analysis of the conditioning channel because they concentrate on a single transistor. Others do not include distortion. On the contrary, [5] does include both noise and distortion and therefore could be used to analyze the power/performance relation of the analog part of the conditioning channel. Still, a dedicated analysis yielding simplified results is presented in sections 4.3, 4.4 and 4.5. It allows straightforward comparison with the analysis on the $\Sigma\Delta$ ADC. In addition, it provides a physical understanding and detailing of the "figures-of-merit" that are introduced next.

4.2 Figures-of-merit

Power/performance relations have been studied on a fundamental level in literature. More empirical, but very practical for comparison of the power/performance ratio achieved in different designs are the so-called FOMs. FOM stands for figure-of-merit. A FOM does not necessarily express a scientific relation between the power and the performance parameters. Rather, it is a means for straightforward comparison of circuit performance by combining a few important specifications in a single number. Some popular FOMs are discussed below.

4.2.1 FOM related to thermal noise

A manifold of variations exist on a FOM that combines power consumption, dynamic range and bandwidth [60]. Often used is:

$$FOM_{DR} \equiv \frac{P}{DR\ BW} \tag{4.1}$$

This FOM follows from thermodynamics. It is typically used for analog circuits where the power consumption is inversely proportional to the noise power. Likewise, it is very applicable to $\Sigma\Delta$ A/D converters because these are normally designed such that thermal noise of the analog input stages dominates over the quantization noise (section 3.2).

The lower limit on this FOM is calculated for a passive network consisting of a single resistor R. Then:

$$DR = \frac{\hat{v}_{\text{IN}}^2}{8kTRBW} \tag{4.2}$$

and the theoretical minimum power consumption equals:

$$P = \frac{\hat{v}_{\text{IN}}^2}{2R} \tag{4.3}$$

such that the FOM equals $4kT$, i.e. 1.6×10^{-20}J. This number is a lower limit and can be used for benchmarking the FOM of active circuits.

4.2.2 FOM including distortion

Often, *SINAD* is used to take distortion and spurious components into account:

$$FOM_{SINAD} \equiv \frac{P}{SINAD\ BW} \tag{4.4}$$

where *SINAD* is expressed as a ratio of powers. This FOM is justified by a derivation further on in this chapter that leads to eq. 4.10 for analog circuits and to eq. 4.16 for ΣΔ ADCs. As such, this FOM can be used for comparison of both types of circuits.

4.2.3 FOM related to signal resolution

For Nyquist A/D converters a FOM related to the number of quantization levels is commonly used. It is expressed in terms of the effective number of bits (*ENOB*) instead of *SINAD*:

$$ENOB = \frac{SINAD_{\text{in dB}} - 1.76}{6.02} \tag{4.5}$$

Furthermore, the minimum of twice the effective resolution bandwidth f_{ERB} and the Nyquist sample rate is used instead of the signal bandwidth. Hence:

$$FOM_{ENOB} \equiv \frac{P}{2^{ENOB}\ \min(2f_{ERB}, f_s)} \tag{4.6}$$

The key difference between the latter and the previous FOMs is in the fact that the power consumption is compared to signal resolution ($\approx 2^{ENOB}$) instead of the ratio of signal *power* over noise and distortion *power* ($\approx 2^{2ENOB}$). The FOM of eq. 4.6 may originate from the fact that for Nyquist A/D converters the hardware complexity (i.e. number of comparators, gain stages, etc.) is proportional to the number of quantization steps [41]. On the contrary, this is not true for a ΣΔ ADC: the relation between the hardware complexity of a ΣΔ ADC and the number of quantization steps is not straightforward. The increase in complexity -and thus power consumption- is a complicated function of the filter order, the over-sample ratio and the number of quantization steps. The effect of the various parameters on *ENOB* is expressed in equation 3.1. The quantization noise power can be reduced by a less-than-linear increase in hardware complexity. Moreover, ΣΔ A/D converters are normally designed such that thermal noise of the analog input stages dominates over the quantization noise anyway (section 3.2).

In general, this FOM can be criticized because it relates power consumption in the numerator to resolution in amplitude (instead of signal/noise *power* in the denominator. *Consequently, it values resolution in amplitude equal to resolution in time, which is conflicting with the Shannon theorem. From a thermodynamics point of view, power consumption of an ADC should be related to its noise power instead of its amplitude resolution.*

When plotting power/bandwidth versus *SINAD* for a large number of published ADCs, the FOM of eq. 4.6 closely predicts both the best-fit trendline and the line connecting the "best"data points; i.e. the ADCs with lowest power/bandwidth for a given *SINAD*. This exercise is reported in [61] for a set of Nyquist and ΣΔ ADCs. Here, it is

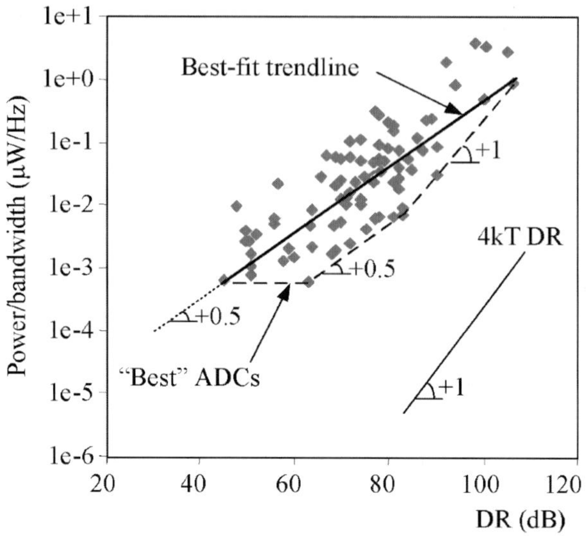

Figure 4.1: *Overview of ΣΔ ADCs as listed in appendix A*

repeated for ΣΔ ADCs only. The data is listed in table A.1 of appendix A and comprises around 70 designs published from 2000 until 2004. The resulting chart of fig. 4.1 proves that, also for ΣΔ ADCs, the FOM of eq. 4.6 closely predicts the power/bandwidth versus *SINAD* trendline. In [61], this anomaly between the theoretical prediction ($\sim 2^{2ENOB}$) and the practical trendline ($\sim 2^{ENOB}$) is attributed to technology constraints, architectural overhead and various implementation related limitations. Here, we add that this trend only shows up when averaging over a lot of designs with different architectures. Within one class of architectures, for example, single-bit ΣΔ ADCs with a feed forward loop filter, a steeper slope is observed. Connecting the "best"designs in fig. 4.1 in a piecewise fashion, instead of using a straight line, yields some additional insights. Converters achieving a high *SINAD* do follow the relation $\sim 2^{2ENOB}$ of eq. 4.4. Likely, thermal noise is determining the power/performance relation of these converters. For converters with a lower *SINAD*, the slope becomes less steep. Possibly, part of the power is consumed in some architectural overhead. That part may not scale -or may scale only weakly- with *SINAD*. Combining this part of the power consumption with contributions that do scale proportionally with *SINAD*, may be the reason why the slope averages and the curve follows $\sim 2^{ENOB}$. Most of the converters with a *SINAD* below 60dB target a reasonably wide bandwidth (see table A.1). Likely, the power consumption is then dominated by bandwidth related limitations (see section 3.4) or by a fixed architectural overhead (e.g. MASH converters). This would explain the flat curve.

For sure, the above explanation remains speculative to some extent. The FOM of eq. 4.6 seems very well suited for predicting the state-of-the-art minimum power consumption that is required to meet a certain performance target. Hence, it is very often used for benchmark purposes. Still, in this book, especially the FOM of eq. 4.4 is used. As motivated, for ΣΔ ADCs, it is meaningful from a physical point-of-view. Therefore, it is well-suited for extrapolation of power consumption of a specific ADC implementation to a varying performance target or vice versa. In addition, in the next section, it is demonstrated that FOM_{SINAD} *is valid both for the ΣΔ ADC as well as for analog circuits. Hence, it allows comparison of the power/performance relation of both. Finally, this FOM can also be used for benchmarking purposes. It yields the same ranking as* FOM_{ENOB} *as long as ADCs of the same bandwidth are compared. This constraint is respected in all benchmarks in this book. For completeness, appendix A includes a table listing both* FOM_{ENOB} *and* FOM_{SINAD} *for a large set of published ΣΔ ADCs.*

4.3 Power consumption in analog conditioning circuits

A relation between current consumption on one hand and thermal noise, intermodulation distortion and signal bandwidth on the other hand is mandatory for a basic analysis of the channel (see chapter 1). This relation is first derived for the analog conditioning circuits. Analog signal conditioning includes amplifying the signal, reducing the signal bandwidth and dynamic range. Hence, filters and amplifiers (having a fixed or a variable gain) are the main building blocks. For simplicity, the analog conditioning circuits are assumed to be based on a differential pair. In practice, this is often the case. In addition, it is in line with the assumption of differential pair based integrator stages in the ΣΔ implementation (page 27). This analogy allows a generalized analysis of the channel and the exchange of analog filter stages for increased A/D performance (see chapter 5). The analog part of the conditioning channel becomes a cascade of differential pairs, each performing a specific function (filtering or amplification). On the other hand, the above assumption excludes specific linearization techniques like a.o. [62], [63], [64], [65]etc.

4.3.1 Power/performance relations

The analysis focuses on the topologies based on a simple differential pair, a differential pair in a global feed-back configuration and a degenerated differential pair as depicted in 4.2.a, b and c respectively. Of course, multi-stage designs can be analyzed in a similar way.

For these topologies, first order relations between the current consumption and noise, distortion and signal level are derived in appendix B. It is assumed that -from all active devices- the input transconductors contribute most to the noise. For the calculations on distortion only the non-linear V/I conversion is included because it is directly related to the quiescent current. Other non-linearities are intrinsic [8]. Their effect can only be reduced by proper transistor dimensioning or circuit design.

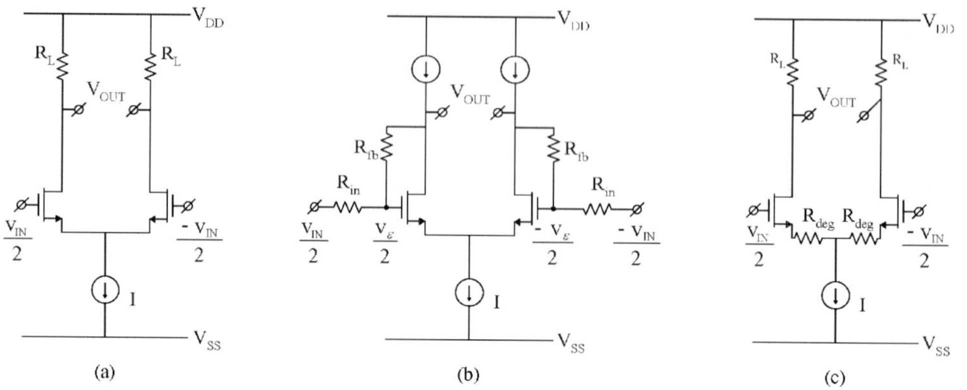

Figure 4.2: *Topologies with differential pair*

Under these conditions, the following power/performance relation is derived in appendix B.

$$I \sim \frac{DR\,BW}{\hat{v}_{IN}^2}\left(\frac{\hat{v}_{IN}}{\sqrt{IM_3}}\right)^{\alpha} \quad (\alpha=0,\ 1\ \text{or}\ 2/3) \tag{4.7}$$

with:

- $\alpha=0$ or 1 for fig. 4.2.a: this assumes the linearity requirements are relaxed such that an open loop configuration becomes possible: the current consumption is either fixed ($\alpha=0$) or the over-drive voltage v_{GT} and quiescent current of the input transistors are adapted to the specification ($\alpha=1$).
- $\alpha=2/3$ for fig. 4.2.b and c: global or local feedback is used in case of challenging linearity requirements. The input transistors are biased near weak-inversion. Hence, the intrinsic linearity of the differential pair is small (i.e. v_{GT} is small) but the linearizing effect of the increased feedback gain (i.e. a large g_m) is dominant because it appears with power 3 in eq. B.7.

The latter implies that, in view of power consumption, it is beneficial to apply feedback instead of relying on the inherent linearity of the input transistors. Therefore, feedback (i.e. an α-value of 2/3) is assumed from now.

4.3.2 Discussion

The above result is commented and compared to literature.

Effect of circuit bandwidth

The presence of BW in eq. 4.7 only refers to the noise bandwidth. In addition to these equations, the quiescent current of all topologies is also proportional to the required circuit bandwidth. Depending on the exact specifications the current required due to noise and distortion (eq. 4.7) or the current required for the circuit bandwidth (proportional to the signal bandwidth) will be dominant.

Results in terms of *SINAD*

Often, circuit performance is specified in terms of *SINAD* (e.g. in data transmission). In cases where distortion and noise contribute in an equal way to the *SINAD* value, eq. 2.2 is valid. Hence, the following relations are assumed:

$$SINAD \sim DR \quad \text{and} \quad SINAD \sim IM_3^{-2} \tag{4.8}$$

Substituting these in eq. 4.7 yields:

$$I \sim \frac{BW\, SINAD^{1+\alpha/4}}{\hat{v}_{\mathrm{IN}}^{2-\alpha}} \tag{4.9}$$

Other applications (like audio and video) have a different tolerance on linearity compared to noise and equation 2.2 does not hold. In the remainder, it is used though in order to simplify the expressions.

It is mentioned here, that for various proportionality relations in this chapter, the units of the left and right hand side differ. This is due to the omission of v_{GT} , kT,etc. By consequence, these relations only express a proportionality between the value at the left and at the right hand side.

Comparison to literature

As discussed in section 4.1, most work on power/performance relations only takes into account *BW* and *DR* [55], [12], [3] and [57]. This corresponds to the results of the $\alpha = 0$ case in the analysis above.

Reference [5] does consider distortion. It presents a.o. a relation for the absolute minimum power of an analog circuit as a function of the required *SINAD* and signal bandwidth. The key assumption in the analysis is that the load of the circuit behaves capacitively over a large part of the signal bandwidth. Hence, this condition is fulfilled for the $\alpha = 2/3$ case in the analysis above because the dominant pole of the open loop transfer function is much smaller than the signal bandwidth.

A further assumption in the referenced paper is the fact that the output swing is rail/rail. Thus, for a fixed gain, V_{DD} can be substituted by $\hat{v}_{\mathrm{IN}}$. The above-derived expression for the current consumption (eq. 4.9) is multiplied with the supply voltage to obtain the power consumption. Next, V_{DD} is substituted by $\hat{v}_{\mathrm{IN}}$, yielding:

$$P \sim \frac{BW\, SINAD^{7/6}}{\hat{v}_{\mathrm{IN}}^{1/3}} \tag{4.10}$$

This is in agreement with the results in [5]. Hence, this comparison serves as a sanity check for the calculations in this chapter.

4.4 Power consumption in a ΣΔ ADC

As mentioned in section 1.2 and further motivated in section 3.2, the present analysis of power consumption concentrates on continuous-time single-bit ΣΔ A/D converters with a feed-forward loop filter. This analysis is conducted for the reference design depicted in fig. 3.4.

4.4.1 Power/performance relations

In [40], it is shown that, in view of an optimal power/performance balance, the ΣΔ ADC should be designed such that the input resistance R_{in} (as defined in fig. 3.4) dominates in the overall noise. Then:

$$DR = \frac{\hat{v}_{\mathrm{IN}}^2/2}{8kTR_{in}BW} \tag{4.11}$$

In the same reference [40], the distortion of the ΣΔ topology of fig. 3.4 is analyzed in terms of HD_3 and assuming the differential pair constitutes the dominant non-linearity. From that, the intermodulation can be derived:

$$IM_3 = \frac{3}{16}\left(\frac{\hat{v}_{\mathrm{IN}}}{v_{\mathrm{GT}}}\right)^2 \frac{1}{(g_m R_{in})^3} \tag{4.12}$$

with v_{GT} the over-drive voltage of the input transistors of the first stage. Note the following:

- the ratio of the signal swing at the input of the transconductor to v_{GT} determines the inherent linearity of the input transistor
- the feedback operation (expressed by the product $g_m R_{in}$) reduces this swing

The latter effect is dominant and therefore lowest distortion is achieved if the input transistor is biased near weak inversion. Finally, the following relation is derived by solving R_{in} from eq. 4.11 and substituting it in eq. 4.12:

$$I \sim \frac{DR\,BW}{{IM_3}^{1/3}\hat{v}_{\mathrm{IN}}^{4/3}} \tag{4.13}$$

The power consumption of other building blocks of the ADC has not been included in this calculation. These blocks have been discussed in section 3.2.2. Their current consumption can be neglected or it depends in the same way on the performance parameters. This is motivated next:

- *the DAC:* It can be implemented using switches only. As CMOS technology has been optimized for switching the associated power consumption is negligible.
- *the quantizer:* It consists of the comparator and some logic gates. Again, the power consumption associated with the switching of a few gates can be neglected. In general the current in the comparator is set by the requirements on the gain-bandwidth product, noise and offset. In the ΣΔ ADC only the BW requirement is important

in view of parasitic loop delay. Other artifacts are less important due to the high loop gain that precedes these error sources. Hence, the current consumption in the quantizer is small.

- *other filter sections:* The higher-order filter sections are normally down-scaled copies of the first stage. Therefore, they obey the same power/performance relation and their contribution in current consumption -although much smaller than that of the first stage- can be accounted for in the value of the proportionality parameter.

4.4.2 Discussion

The mathematical relations are commented, rephrased and compared to popular FOMs for ΣΔ ADCs.

Effect of circuit bandwidth

The above analysis is valid as long as the current consumption is set by noise and distortion requirements. In case sample rate associated limitations (see section 3.4) become dominant these will further increase current consumption.

Although BW explicitly appears in eq. 4.13 it actually only refers to the noise bandwidth as introduced in eq. 4.11. The bandwidth requirements on the sub-circuits of the ADC are not truly taken into account. For most ΣΔ ADCs, the power is indeed determined by noise and distortion instead of bandwidth requirements. Hence, eq. 4.13 is used to extrapolate the power consumption of most continuous-time ΣΔ ADCs.

Results in terms of *SINAD*

Again, equation 4.13 can be expressed in terms of *SINAD* assuming noise and distortion contribute in a fixed ratio. Eq. 4.8 is substituted in 4.13 yielding:

$$I \sim \frac{SINAD^{7/6} BW}{\hat{v}_{IN}^{4/3}} \tag{4.14}$$

This result is the same as that for the analog circuits using feedback.

Comparison with popular FOMs

In section 4.2 it was motivated that for design purposes the FOMs in terms of *DR* and *SINAD* (eq. 4.1 and 4.4 resp.) are more meaningful for ΣΔ ADCs than the FOM in terms of quantization steps (eq.4.6). These FOMs are now compared with the derived analytical relations.

First, the popular FOM of eq. 4.1 (including *DR* only) is considered. It is compared to the following modified FOM derived from eq. 4.13:

$$FOM = \frac{P}{DR\;BW} \; \frac{\hat{v}_{IN}^{4/3}\; IM_3^{1/3}}{V_{DD}} \tag{4.15}$$

It expresses the same linear relation between power consumption of the ADC and the product of bandwidth and dynamic range. In addition, it shows the way to including the effect of intermodulation distortion and the effect of $\hat{v}_{\text{IN}}$ and the supply voltage V_{DD}. Notice, the dimension of this FOM deviates from that of all other FOMs. This FOM is used a.o. in chapter 8 to compare the performance of an analog VGA to that of a ΣΔ ADC.

Secondly, eq. 4.14 is used to motivate the use of FOM_{SINAD} (eq. 4.4) throughout the book. The analytically derived eq. 4.14 yields the following modification as compared to eq. 4.4:

$$FOM = \frac{P}{SINAD^{7/6}\ BW}\ \frac{\hat{v}_{\text{IN}}^{4/3}}{V_{DD}} \tag{4.16}$$

The modification compared to the conventional FOM_{SINAD} is three-fold:

- $\hat{v}_{\text{IN}}$ *is taken into account:* A fixed *DR* is achieved at less power as the input signal becomes larger: power decreases quadratically with $\hat{v}_{\text{IN}}$. The opposite is true for distortion: for a fixed IM_3-value g_m must increase almost proportionally with $\hat{v}_{\text{IN}}$ (eq. 4.12). The combined effect results in an -almost- linear relation between the FOM and $\hat{v}_{\text{IN}}$.

- *the FOM is normalized to the supply voltage:* In fact, this eliminates V_{DD} from the FOM leaving the quiescent current -instead of the power consumption- in the numerator. Indeed, in the above analysis performance was assumed independent of V_{DD} and only related to the bias current[1] .

- *SINAD is included more accurately:* because of distortion, *SINAD* increases more than proportionally with power consumption.

Furthermore, the following remarks on the application of the modified FOMs hold:

- *modified FOMs and technology scaling:* The normalization to V_{DD} eliminates the dominant effect of technology scaling on the FOM. Therefore, these FOMs especially express the cleverness in exploiting the technology or in dealing with it. Still, smaller effects of technology scaling, like the increasing g_m/I or the decreasing inherent linearity of the transistor, remain included.

- *modified FOMs and different ΣΔ topologies:* Notice that the same (simplified) FOMs would have been derived for any topology different from fig. 3.4 as long as a differential pair constitutes the dominant power consumption.

It would have been interesting to compare some published ΣΔ ADCs according to FOM_{DR} and the suggested modification thereof. Unfortunately, many publications do not mention the input voltage. Therefore, important references could not be included and the comparison becomes less meaningful. Still, this exercise is conducted for a few ΣΔ designs published by Philips Research for which all required data is internally available. Table 4.1 ranks the publications according to the modified FOM (eq. 4.15) in the utmost right column. From this table a few observations are made:

[1] This assumption is valid as long as the ADC is easily scalable. This seems to be the case for the implementations in chapter 8 for the next CMOS generations to come.

Table 4.1: *Comparison of conventional and modified DR-based FOM*

Reference	Technology	FOM_{DR} eq. 4.1	$FOM_{DR,modif}$ eq. 4.15
[30] CDMA	0.18μm CMOS	$4 \times 10^{-18} J$	$1 \times 10^{-18} J$
[30] UMTS	0.18μm CMOS	$1 \times 10^{-17} J$	$6 \times 10^{-18} J$
[30] GSM	0.18μm CMOS	$2 \times 10^{-17} J$	$6 \times 10^{-19} J$
[38] GSM	0.35μm CMOS	$5 \times 10^{-17} J$	$2 \times 10^{-18} J$
[32] Bluetooth	0.18μm CMOS	$1 \times 10^{-16} J$	$1 \times 10^{-18} J$
[2] AM	0.25μm CMOS	$2 \times 10^{-16} J$	$2 \times 10^{-18} J$
[2] FM	0.25μm CMOS	$3 \times 10^{-16} J$	$2 \times 10^{-18} J$
[7] speech	0.5μm CMOS	$6 \times 10^{-16} J$	$4 \times 10^{-19} J$

- the modification to the FOM reduces the variation in the achieved values because the influence of V_{DD} is eliminated: $FOM_{DR,modif}$ varies by a factor of 10 while for FOM_{DR} the ratio of the best to the worst FOM is about 100.
- the ranking especially differs for designs with a rather small input voltage: for example [7] with only $56mV$ of input signal is ranked higher -thus fairer- using the modified FOM.

Concluding the discussion on the power/performance analysis of the $\Sigma\Delta$ ADC the following use of FOMs is advised:

- for design purposes the elaborated FOMS of eq. 4.15 and eq. 4.16 can best be used since they express a detailed relation between the various performance parameters on the power consumption.
- for bench-mark purposes the conventional FOMs of eq. 4.1 and eq. 4.4 are easier because they need less published data.

4.5 Power consumption in digital conditioning circuits

The functionality of the digital conditioning circuits is in:

1. decimating the incoming data, implying:
   - the filtering of noise and interferers outside the wanted channel

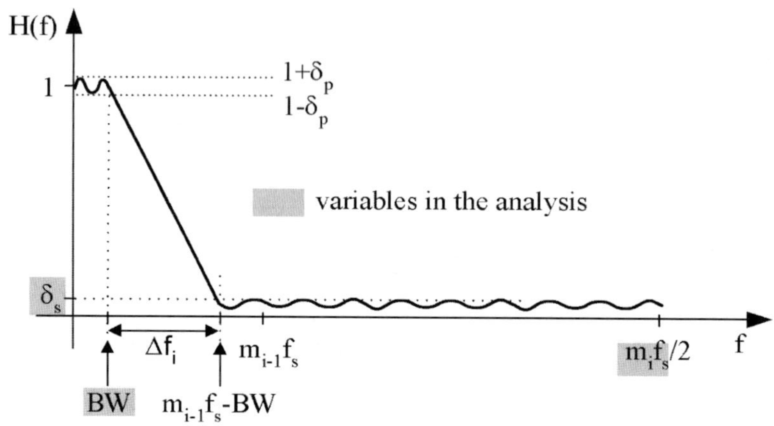

Figure 4.3: *Definition of the decimation filter characteristics*

- the reduction of the high sample rate to the Nyquist sample rate.

2. reducing the dynamic range by scaling the word-length (digital VGA).

Both aspects are discussed next.

Decimation filtering: The decimation function is clarified based on fig. 4.3 depicting the transfer function of the ith decimation filter (assuming multi-stage decimation). It provides suppression of noise (shaped quantization errors of the ΣΔ ADC) and interferers (remaining in case of weak analog filtering) that are present around the target sample rate $m_{i-1}f_s$. Next, the filtered signal can be re-sampled. The key specifications on the filter are:

- *the transition bandwidth* Δf*:* in the ith decimation stage it equals $\Delta f = m_{i-1}f_s - 2BW$. In the last decimation stage the transition band equals $\Delta f = f_s - 2BW$.
- *the stop-band suppression* δ_s*:* In a worst-case scenario all noise or interferer energy can be present in a single tone. Therefore, a suppression equaling the *SINAD* specification is required.
- *the pass-band ripple* δ_p

In the power/performance analysis only the stop-band suppression is included as a variable. The pass-band ripple and the transition band in the last decimation stage are assumed to be fixed by the application[2] and do not take part in the balancing between analog and

[2] In digital communications for instance, the Nyquist sample rate f_s is set by the symbol, bit or chip rate (e.g. a $3.82MS/s$ chip rate for UMTS). The signal bandwidth BW is determined by the modulation index or by the requirement of a guard band between neighboring channels. A different example is found in digital audio systems. Here, the Nyquist sample rate is fixed for standardization reasons (e.g. $f_s = 44.1kS/s$ for digital audio) while the signal bandwidth can be limited to the audible frequency range.

digital conditioning. In addition, the number of decimation stages and the decimation factor are less important because these belong to the digital optimization. *The analysis assumes an optimal digital design and derives a power/performance relation that can be used for extrapolation of the power consumption of this design to varying performance requirements or vice versa.*

Scaling: The second function is in reducing the dynamic range of the received signal to the *SINAD* that is required for further processing, like for example digital demodulation. In digital, this can be implemented in a rather straightforward way by simple bit-shifting if a control signal (a Received Signal Strength Indication or RSSI) is available. Therefore, it is not taken into account in the calculations. Instead, the decimation filtering is emphasized.

4.5.1 Filter functions

Digital filters can be divided into two main classes: the finite impulse responses (FIR) and the infinite impulse response filters (IIR). Although IIR filters can realize a sharp transition band at low complexity compared to FIR filters they are prone to coefficient growth and stability problems. In addition, IIR filters suffer from phase distortion. This is a major drawback obstructing the use of IIR filters in many applications.

Below, some often-used filter functions that are inherently linear-phase -or can be designed as such- are briefly discussed. More detail can be found in [25], [66], [67], [68], a.o.

CIC filters [69], [70]: Cascaded-Integrator-Comb filters consist of an integrator section at the high sample rate and a comb section operating at the low sample rate. They realize a frequency response:

$$H(f) = \left| \frac{\sin(\pi f m T)}{m \sin(\pi f T)} \right|^k \quad \text{with } T = \frac{1}{m f_s} \tag{4.17}$$

k is the number of cascaded sections, m is the decimation factor. This frequency response has zeros around multiples of the decimated sample rate. Hence, CIC filters efficiently suppress the frequency bands that would fold back into the wanted channel when resampling. They are often used in a first decimation step because they don't need multiplying with coefficients but just a single addition per tap is made. Disadvantages are in the restrictions on the pass-band width (it must be small compared to the decimated sample rate to keep droop limited) and on the weak roll-off outside the pass-band. This limits their application to decimation until four times the Nyquist sample rate.

Half-band filters [71]: Linear-phase half-band filters are characterized by a symmetrical transition band limiting their use to decimation by a factor of 2. Because of the symmetry, they can be implemented with half the number of multiplications compared to arbitrary filter designs. Hence, a half-band design is very attractive for use in the first decimation stage running at the highest frequency.

Generic FIR filters: Especially in the final decimation stage a sharp transition band and droop-correction need to be realized. Typically, this last stage needs the largest number of taps. On the other hand, it operates at the lowest frequency making the switching dissipation acceptable. A generic FIR design can realize these requirements.

4.5.2 Power/performance relations

In general, the power consumption P of any digital block is determined by the supply voltage V_{DD}, the total capacitance C that needs to be switched, the clock frequency f and the parameter α expressing the average activity of the gates:

$$P = \alpha f C V_{DD}^2 \tag{4.18}$$

The aim of the analysis is in comparing power consumption of analog and digital signal conditioning. *Therefore, a relation between the power consumption in the decimation filter and the parameters ENOB, mf_s and f_s is mandatory*[3]. Contributions that do not scale with these parameters can be disregarded. In addition, the activity α and the supply voltage V_{DD} in eq. (4.18) are considered independent thereof[4]. Hence, *the prime challenge is in expressing the capacitance C (dominated by the number of gates) in terms of the quoted parameters.* This is done in appendix C. In addition, the following upper limit on the power/performance relation for digital decimation filters is derived there:

$$I_{worst-case} \sim O\left(\frac{ENOB^3}{\frac{\Delta f}{2f_s}} mf_s \ln m\right) \tag{4.19}$$

ENOB appears in a cubic relation: in a worst-case scenario, the number of filter taps, the word-length of the data and the number of computations (in terms of partial additions) are all proportional to *ENOB*. In the appendix, it is shown that for the first decimation stages and for CIC implementations, a quadratic relation is more likely. Furthermore, the highest sample rate mf_s and the smallest transition band $\Delta f/2f_s$ are assumed, while the factor $\ln m$ is a measure for the number of consecutive decimation stages. Further on, it is neglected because its influence is small.

4.5.3 Discussion

Again, the above results are further commented in the sense that the major influence of the sample rate is highlighted. Also, the benefit of technology scaling is drawn to the attention.

[3] *ENOB* is used here instead of *SINAD* in the discussion on analog. For simplicity, it is assumed that all bits are *effective*; i.e. we do not distinguish between the effective-number-of-bits (*ENOB*) and number-of-bits (*NOB*).

[4] The average activity is determined by the input data and the implementation of the filter. The supply voltage is set by the technology. Sometimes, though, it is adapted according to the required processing speed. This potential dependence on f_s is disregarded in the present analysis. It can easily be taken into account afterwards.

Effect of sample rate

The derived results are valid if power consumption of the digital conditioning channel is dominated by the decimation filters. In many digital designs, other contributions are important as well. These are typically due to parasitics, e.g. wiring capacitance or due to additional circuits, e.g. for test or control. In addition, the power consumed in the clock distribution can be very important. Notice that this contribution scales with mf_s.

In low power digital design it is common practice to adapt the supply voltage according to the required sample rate[5] . The variation of the supply voltage is limited by the noise margin on the low side and by reliability issues on the high side. Within this window, the supply voltage can be scaled to the sample rate. Using eq. 4.18, power consumption then becomes cubic in mf_s instead of proportional. In the above analysis, this variation has been disregarded for simplicity reasons.

The various remarks in this paragraph do underline the prime importance of the sample rate on the power consumption of a digital block.

Technology scaling

Any analysis of power consumption in digital blocks would not be complete without stressing the benefits of technology scaling. While for analog circuits technology scaling hardly brings any power savings (in fact, beyond 0.25μm-CMOS a penalty in power consumption is more likely [5]), for digital circuits the power decrease is substantial. Porting the decimation filter from one technology generation to the next reduces the total capacitance by a factor s_{tech}, with s_{tech} being the technology-scaling factor [72]:

- the area scales by $1/s_{tech}^2$
- the gate-oxide thickness scales by $1/s_{tech}$

Hence, the current consumption of the digital block scales accordingly:

$$I_{digital} \longrightarrow I_{digital} \cdot s_{tech} \qquad (s_{tech} \text{ is the technology scaling factor}) \tag{4.20}$$

In addition, the supply voltage can be scaled by a factor s_{supply}. According to equation 4.18 this results in the following power saving:

$$P_{digital} \longrightarrow P_{digital} \cdot s_{tech} \cdot s_{supply}^2 \qquad (s_{supply} \text{ is the voltage scaling factor}) \tag{4.21}$$

Assuming s_{tech} and s_{supply} both approximate 0.7[6], per transition, some 65% of power consumption is saved going from one technology generation to the next. From Moore's Law, it can then be derived that the dynamic power consumption of digital circuits is reduced by a factor of 1000 every 10 years; i.e. in about 6 to 7 technology generations.

[5] In fact, this technique can be actively applied to lower the power consumption of a digital design: using parallelism in the implementation allows operation at a lower sample frequency and thus at a lower supply voltage. The net effect is a decreased power consumption.

[6] This has been true in the past but according to the ITRS roadmap, supply voltage scaling may be slowing down in view of transition speeds.

4.6 Comparison

The power/performance relations derived for the major building blocks of the channel are repeated here and discussed afterwards.
For analog conditioning circuits eq. 4.7 is valid:

$$I \sim \frac{BW\, SINAD^{1+\alpha/4}}{\hat{v}_{\mathrm{IN}}^{2-\alpha}} \qquad (\alpha = 0,\ 1 \text{ or } 2/3)$$

Especially the $\alpha = 2/3$ case is interesting: it is valid for circuits using local or global feedback.
For the $\Sigma\Delta$ ADC eq. 4.14 is repeated:

$$I \sim \frac{BW\, SINAD^{7/6}}{\hat{v}_{\mathrm{IN}}^{4/3}}$$

For the decimation filter eq. 4.19 has been derived (neglecting the small influence of the logarithmic function):

$$I_{worst-case} \sim O\left(\frac{ENOB^3}{\frac{\Delta f}{2f_s}}\, mf_s \ln m\right)$$

Notice, the results are quoted in terms of *SINAD* or *ENOB* (instead of IM_3 and *DR*) to allow easy comparison between analog and digital.

Comparison of the power/performance relations

The power/performance relation of the $\Sigma\Delta$ ADC resembles that of analog conditioning circuits. Below, some minor differences are explained. Next, these power/performance relations are compared to the power/performance relation of digital conditioning circuits, leading to some generic thoughts on the digitization of the conditioning channel.

Power/performance relation for $\Sigma\Delta$ A/D conversion versus that for analog conditioning: The power/performance relation of the $\Sigma\Delta$ ADC is the same as that of the analog conditioning circuits with $\alpha = 2/3$. Therefore, the FOMs of eq. 4.15 through eq. 4.16 that were derived for a $\Sigma\Delta$ ADC, are applicable to analog circuits as well. This results from the assumption that the ADC power consumption is dominated by its analog circuits and that these consist of similar differential pairs as those used for the analog conditioning circuits. However, the proportionality factor between the left and right-hand side of the above relations differs for two reasons:

- *Different linearity:* The $\Sigma\Delta$ ADC feeds back a pulse-density modulated signal instead of a simple analog signal. Therefore, distortion in the input stage of the $\Sigma\Delta$ ADC is twice larger than that of the analog circuits (eq. B.8 versus eq. 4.12). This is detailed mathematically in appendix D.

- *Contribution of cascaded stages:* The ΣΔ ADC as well as the analog conditioning part consists of a cascade of sections all adding to the overall current consumption. In both cases, the consecutive stages can have increasing noise and distortion because of the preceding gain. This is especially true for the ΣΔ ADC where the loop filter can have very high gain at low-frequencies. Because of the closed-loop operation, the filter only needs to meet a gain/phase criterion at high frequencies. On the contrary, the design of the analog cascade is often much more constrained. For instance, the transfer function is tightly specified and, in the likely case of open-loop operation, the gain of the consecutive stages is limited. Because of these constraints, a cascade of several analog sections is likely to consume more than the loop filter of a ΣΔ ADC of the same order.

The latter observation is supported by the example of an FM receiver in chapter 8. Table 8.5 compares the power consumption of a highly analog conditioning channel (first column) to that of a the ADC in a highly digital channel (last column).

Power/performance relation of digital conditioning versus that for analog conditioning or ΣΔ A/D conversion: Power consumption in the decimation filter is proportional to *ENOB* or a power of *ENOB* while in analog circuits and ΣΔ ADCs it is an exponential function of *ENOB* (i.e. linear in *SINAD*). Even under the worst-case assumption of eq. 4.19 the power increase as a function of *ENOB* is moderate compared to that in analog circuits.

Example: conditioning channel for FM receiver (section 8.3)
In fig. 4.4 the power/performance relations are applied to an example conditioning channel. In the referenced implementation, the conditioning channel achieves 14 bits of resolution and the power consumption of the ΣΔ ADC and that of the digital decimation filter are about equal. From this measurement point, the power consumption is extrapolated as a function of ENOB. In this example, P_{analog} *represents the power consumption of the ΣΔ ADC.* $P_{digital}$ *is extrapolated both as a square and as a cubic function of ENOB in order to demonstrate the minor difference.*

In general, equations 4.9, C.6 and C.7 can only be used for extrapolation purposes because they express a proportionality. Then, the crossing of the power/performance curves depends on the actual implementation and on the various parameters in eq. 4.9, C.6 and C.7.

Conclusions with respect to digitization: The power consumption of the complete channel equals the sum of the analog (or ADC) power and the digital power consumption and, in fig. 4.4, is represented by the dotted envelope curve. From this example, it can be generalized that:

- in channels targeting a low *ENOB*-value, the power consumption of digital blocks is likely to be dominant

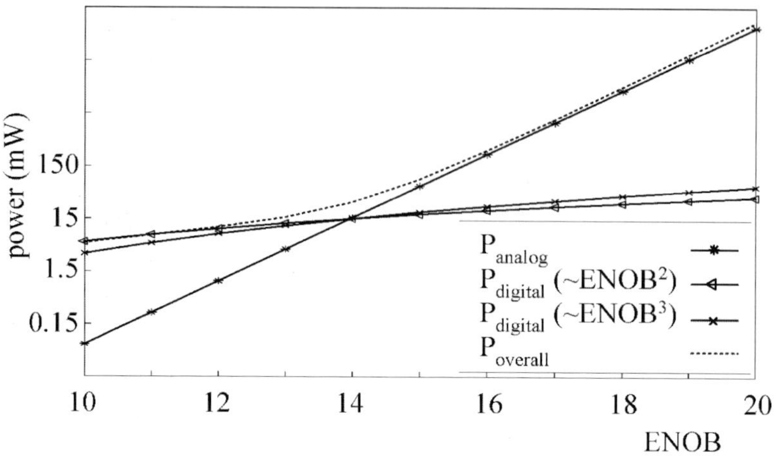

Figure 4.4: *Power consumption of analog and digital blocks as a function of ENOB (normalized at 14 bits corresponding to an example in section 8.3)*

- in channels targeting a high *ENOB*-value, the power consumption is dominated by the analog blocks

In a generalized version of fig. 4.4, P_{analog} would either represent the power consumption of the analog circuits (in case of a highly analog channel) or it is the power consumed in the ΣΔ ADC (in case of a highly digitized channel). It is not self-evident that digitization of the channel implies a power reduction. On one hand, this may be the case because:

- while the same power/performance *relation* is valid for analog circuits and the ΣΔ ADC, the *absolute* power consumption of a high-order ΣΔ ADC, likely, is lower than that of a cascade of several analog circuits (see discussion on page 53 on cascaded stages)

On the other hand, though, digitization may just as well lead to a huge power increase for the ADC:

- due to the high resolution and bandwidth demands on the ADC, the limitations of section 3.4 are hit and its power consumption increases faster than what is predicted by eq. 4.14

In addition, the power consumption of the digital part increases. The power consequences of digitization are further explored in chapters 5 and 6 for various conditioning channels. *In general, digitization of the conditioning channel does not necessarily imply a power saving.*

Finally, considering:

- the ever-increasing resolution demand, i.e. a higher *ENOB* target

- the benefits of technology scaling, i.e. the $P_{digital}$-curve shifts down

it can be concluded that:

- power consumption of analog blocks becomes dominant in more and more conditioning channels
- this is also true in highly digitized channels: there, the "analog" power is consumed in the $\Sigma\Delta$ ADC

These observations underline the growing, paramount importance of low-power analog circuit design. However, the power/performance relation of *analog circuits only improves gradually* in time. In order to safeguard an acceptable power consumption for high-resolution conditioning channels in advanced technologies, *architectural innovation is needed.* Current research approaches on the architectural level include:

- digitally tuning/assisting analog circuits [20]
- dynamic scaling of power consumption to the instantaneous performance target
- reducing the number of analog stages (i.e. the number of current branches) by merging functionality

Especially the latter two techniques are illustrated in chapter 6, introducing conditioning $\Sigma\Delta$ ADCs, and in the various designs presented throughout chapters 7 and 8.

Comparison to literature

The curves of fig. 4.4 correspond to the linear and logarithmic relation as a function of *SINAD* derived by a.o. Vittoz [55] and Meindl [56] for power consumption in analog and digital respectively. However:

- the referenced papers concentrate on *fundamental limits* for power consumption of *a single analog or digital cell*

while here:

- far more elaborated *relations for the estimated consumption* of *actual channels* are derived;
- in addition, the derived relations are used further on to improve the overall power/performance ratio of the channel crossing the traditional boundaries between analog, mixed-signal and digital design.

4.7 Conclusions

Because of the lack of analytical power/performance relations for analog and digital conditioning blocks in literature, a dedicated analysis is performed in this chapter. This analysis results in the observations listed next.

The power consumption of a single-bit continuous-time ΣΔ ADC can be extrapolated to a modified performance target, using the following Figure-of-Merit:

$$FOM = \frac{P}{SINAD^{7/6}\ BW} \frac{\hat{v}_{\mathrm{IN}}^{4/3}}{V_{DD}}$$

On the other hand, for ranking purposes, more common FOMs can be used because they require less parameters.

It is shown that the power/performance *relation* of analog circuits is similar to that of a continuous-time ΣΔ ADC. The *absolute value* of the power consumption of a high-order ΣΔ ADC is likely to be lower than that of a cascade of several analog circuits because of the overall feedback.

Consequently, digitization of the channel -where the analog circuits are replaced by a difficult ADC and consecutive digital conditioning- does not self-evidently yield a power saving, nor does it necessarily yield a power penalty.

The power consumption of a decimation filter is only a weak function of the *ENOB* specification (when compared to the relation for an analog function) but it is strongly affected by the sample rate.

The power consumption of analog circuits or of the A/D converter becomes dominant in most conditioning channels. This is also true in highly digitized channels.

A power reduction in the conditioning channel requires architectural innovation of analog blocks or of the ΣΔ ADC.

These observations are used further on for exchanging analog and digital conditioning in a channel.

Chapter 5

Full-analog and full-digital conditioning channels

Building on the results of the previous chapter, two channel architectures are compared in terms of their power/performance ratio. The channels differ in the degree of digitization of the signal conditioning[1]*. The first architecture is a conventional channel with full-analog signal conditioning. The second channel architecture has full-digital signal conditioning and therefore is highly flexible. The full-digital conditioning channel requires multi-channel A/D conversion at a very high sample rate. It is shown that, depending on the application, a full-digital conditioning channel may not be feasible in view of current* $\Sigma\Delta$ *ADC performance and with the present technologies. Especially, the requirement for a very high sample rate is demanding, both for the* $\Sigma\Delta$ *ADC and for the digital part. Hence, alternative channel topologies are required in order to maintain a competitive power/performance ratio while pursuing digitization. This is the topic of chapter 6.*

5.1 Full-analog conditioning channel

The operation of the channels (in this and the next chapter) is clarified with the example input signal of fig. 5.1.a. It consists of a small wanted channel, a slightly stronger adjacent channel and two much stronger far-off interferer channels that could cause intermodulation components in the wanted channel. At the output of each conditioning channel, only the digitized wanted channel is available with a specified *SINAD* and sample rate (fig. 5.1.b). As a starting point, a channel with full-analog conditioning is discussed. Further on it serves as a benchmark for the power/performance ratio of digitized channels.

[1]The term "signal conditioning" is used here in the limited sense; i.e. it only refers to analog/digital channel filtering and analog VGA or digital word-length scaling (see nomenclature defined in fig 1.3)

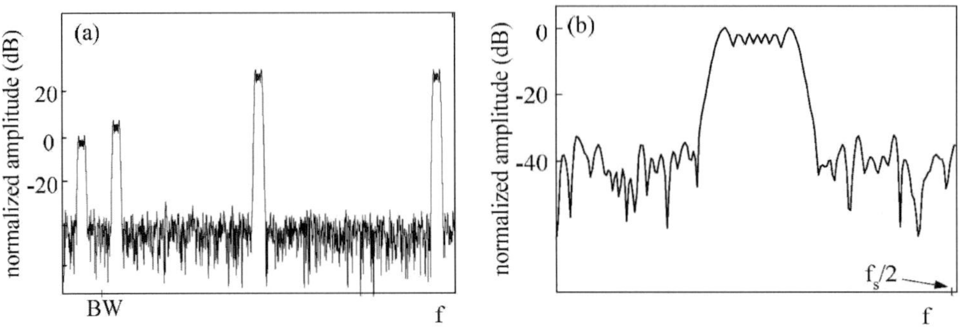

Figure 5.1: *Example input spectrum (a) and output spectrum (b) of the conditioning channel*

5.1.1 The conditioning channel

Fig. 5.2 depicts a block diagram of a conventional, highly-analog conditioning channel with $\Sigma\Delta$ A/D conversion. The bottom graph of fig. 5.2 shows the conditioning of the signal in the various stages. The parameters and symbols that are used here have been defined in section 2.5. DR_i indicates the value of the input-referred dynamic range for the wanted channel. $SINAD_o$ is the value of the *SINAD* in the bandwidth of the wanted channel available at the output of the conditioning channel. If the subscript is omitted, it is explicitly specified to which bandwidth and position in the channel *DR* and *SINAD* refer.

- *the channel filter* $H_{ch}(s)$*:* reduces the signal bandwidth by a factor p resulting in a p-times lower integrated noise power and suppresses the large interferers with overall[2] rms-value $q\ v_{\mathrm{MAX}}$

- *the VGA:* reduces the amplitude range of the input signal by varying the gain between a maximum (G_{MAX}) and a minimum value (G_{MIN}). This is typically allowed in a receiver-type of application. For simplicity, the output-referred noise is assumed to be the same in all gain settings in the remainder.

- *the* $\Sigma\Delta$ *ADC:* digitizes the signal adding a lot of quantization noise and increasing the bandwidth requirements due to the over-sampling.

- *the decimation filter* $H_{dec}[z]$*:* suppresses the shaped noise and reduces the sample rate such that at its output the specified $SINAD_o$ and signal bandwidth become available.

Clearly, the filter $H_{ch}(s)$ is the most challenging block in this architecture: it must achieve low noise and distortion in the presence of a wide-band input signal with a large amplitude range. The requirements on the subsequent blocks become more and more relaxed.

[2]It is assumed that the overall amplitude of the sum of all interferers is a factor q larger than that of the maximum wanted channel. If this is not the case then q should be put to 1.

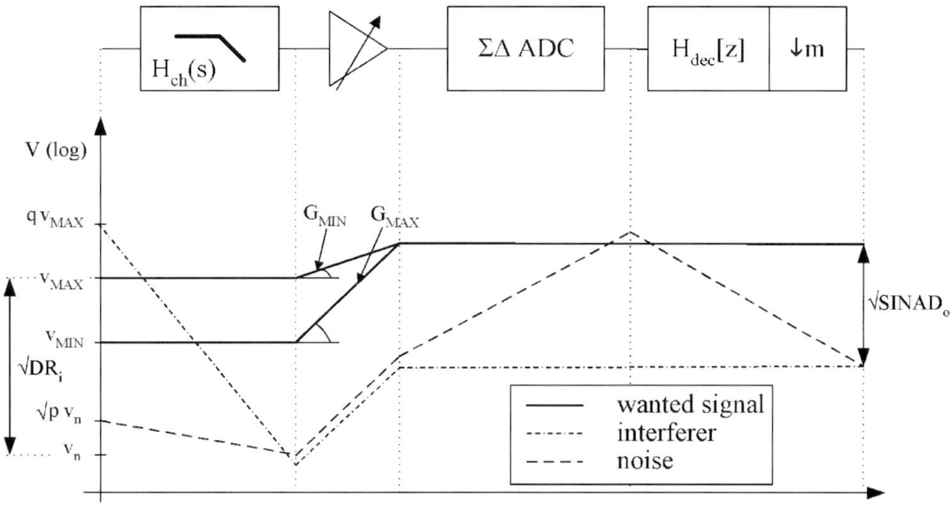

Figure 5.2: *A conventional conditioning channel with ΣΔ ADC*

Not visible in this diagram are the bandwidth requirements on the analog circuits throughout the channel. The bandwidth of the filter circuits must often be large compared to the entire signal bandwidth (including interferers). The bandwidth of the analog circuits in the ΣΔ ADC is determined by the sample frequency $m_1 f_s$.

An example implementation of a conventional conditioning channel for a FM receiver is presented in chapter 8.

5.2 Full-digital conditioning channel

As a counterpart to the full-analog channel, multi-channel A/D conversion and full-digital signal conditioning are analyzed next.

5.2.1 The conditioning channel

The full-digital conditioning channel of fig. 5.3 offers great flexibility (to accommodate changing standards or multi-channel and multi-mode operation) and allows easy technology scaling (to shrink the area or power consumption). The burden of this solution is on the ΣΔ ADC. Both the interferers -in adjacent or in far-off channels- and the wanted channel are digitized (fig. 5.4). Because of its inherent alias-filtering, the continuous-time ΣΔ ADC is robust to high-frequency products at its input. For example, these may have been generated by a mixer that precedes the channel of fig. 5.3. Table 5.1 compares the requirements on the multi-channel ΣΔ ADC to those on the ΣΔ ADC in the full-analog channel. The multi-channel ADC needs to convert a p-times larger bandwidth. Its stable input range must be dimensioned to the amplitude of the dominant interferer, even while the wanted channel may be small due to the lack of preceding gain. As such, both a high

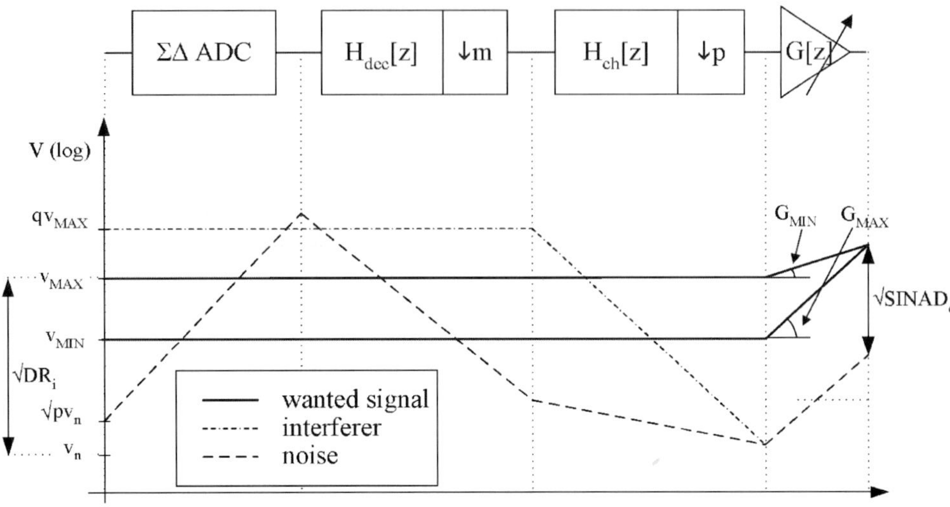

Figure 5.3: *Full-digital conditioning channel with multi-channel ΣΔ A/D conversion*

Table 5.1: *Specifications on the ΣΔ ADC in the full-analog and in the full-digital channel*

Specification	ΣΔ ADC in full-analog channel	ΣΔ ADC in full-digital channel
SINAD	$SINAD_o$	$q^2 DR_i / p$
bandwidth	BW	$p\,BW$
maximum input	$G_{\mathrm{MIN}}\ v_{\mathrm{MAX}}$	$q\ v_{\mathrm{MAX}}$
intermodulation[a]	$2/\sqrt{SINAD_o}$	$2/q\sqrt{DR_i}$

[a]referred to the maximum input signal of the ADC, using eq. 2.2

IM_3 is necessary because of the large interferers and the noise must remain low compared to the smallest wanted signal. Note that these strict requirements are not caused by the wanted channel but are due to the strong interferers. The above requirements result in a more than p-times larger sample rate: next to the higher conversion bandwidth, it is very likely that a higher over-sample ratio and loop filter order are mandatory for the ADC to meet the signal-to-quantization-noise ratio. Possibly, the specifications on the ΣΔ ADC rise above what is feasible in view of state-of-the-art performance (see limitations in section 4.4).

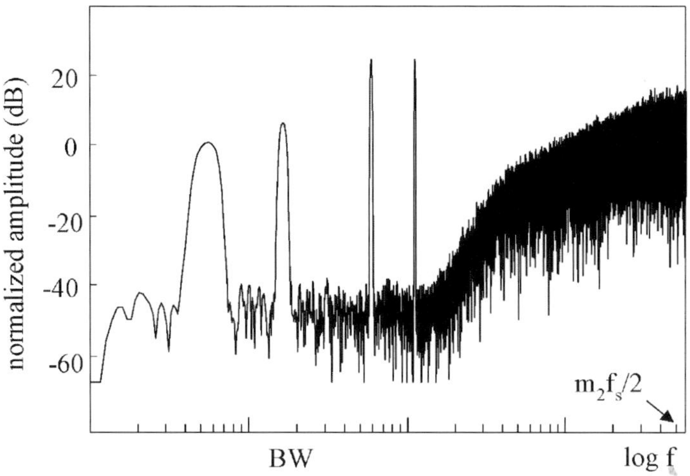

Figure 5.4: *Output spectrum of the multi-channel ΣΔ ADC in the full-digital channel*

Finally, the channel selection is performed in the digital domain in the filter $H_{ch}[z]$ and the word length can be scaled in the block $G[z]$ (i.e. digital VGA). In practice, $H_{ch}[z]$ and $G[z]$ are merged with the decimation filter $H_{dec}[z]$.

5.2.2 Power/performance analysis

First, the power consumption of the multi-channel ΣΔ ADC is compared to that of the ΣΔ ADC in the full-analog channel. Next, the overall power consumption of the full-digital channel is analyzed relative to that of the full-analog channel (fig. 5.2). For completeness, the requirements on peripheral circuits (e.g. reference circuits) in both channels are briefly compared as well.

The ΣΔ ADC

The multi-channel ΣΔ ADC is compared to the simple ΣΔ ADC in the full-analog channel. The analysis is conducted from two angles being resolution (i.e. noise and distortion) and bandwidth requirements.

Noise and distortion: The power/performance relation of eq. 4.13 and the requirements specified in table 5.1 are used to compare the current consumption I_{ADC}(digital CCh) of the ΣΔ ADC in the full-digital channel to the current I_{ADC}(analog CCh) of the ΣΔ ADC in the full-analog channel. The abbreviation CCh indicates "conditioning channel":

$$\frac{I_{\text{ADC}}(\text{digital CCh})}{I_{\text{ADC}}(\text{analog CCh})} = q\left(\frac{DR_i}{SINAD_o}\right)^{7/6} G_{\text{MIN}}^{4/3} \tag{5.1}$$

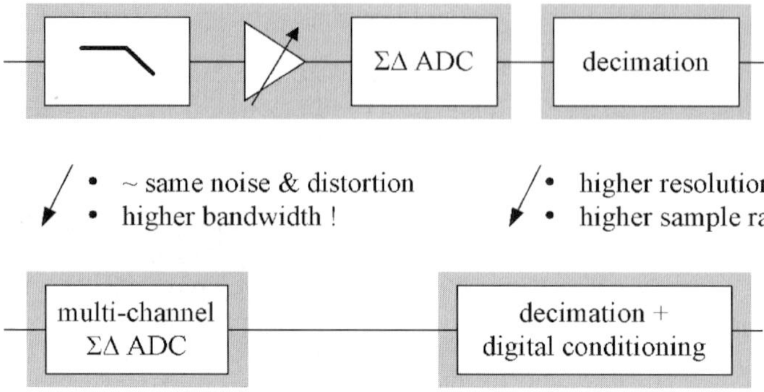

Figure 5.5: *Comparison of a full-analog and a full-digital conditioning channel in terms of power/performance*

The lack of analog filters or VGA results in a very large resolution requirement on the ΣΔ ADC and a consequent power increase. Especially in wireless communication systems, the difference between DR_i and $SINAD_o$ can be very large. A further increase is due to the q-times larger overall amplitude because of the interferers while the wanted signal is G_{MIN} times smaller.

Bandwidth: The multi-channel ΣΔ ADC requires a much larger circuit bandwidth:

- the bandwidth of interest is p times larger
- the over-sampling factor m_2 must be higher in order to achieve the required $SQNR$

From both angles, the multi-channel ΣΔ ADC is significantly more difficult -and therefore more power-hungry- than the ΣΔ ADC in the conventional channel.

The conditioning channel

The above power increase for the ΣΔ ADC should be put into perspective (see section 4.6). In fact, the power consumption of the overall channel should be compared to that of the conventional channel instead of only comparing the ADCs. This analysis is split into two steps (see fig. 5.5). First, the power consumption of the multi-channel ΣΔ ADC is compared to the joint power consumption of the analog part of the conventional channel. Next, the digital part of both channels is compared.

Analog part: The multi-channel ΣΔ ADC is compared to the cascade of analog circuits and ΣΔ ADC in the conventional channel. On one hand, noise and distortion specifications on the multi-channel ΣΔ ADC are the same as those on the input stage -i.e. the filter- of the conventional channel. On the other hand, the power/performance dependence of the multi-channel ΣΔ ADC is similar to that of the input stage of the conventional channel (assuming the input transconductor dominates performance, see chapter 4). Therefore, one can argue that, power-wise, the ΣΔ ADC in the full-digital architecture is competitive to the input stage of the conventional channel.

Moreover, in the discussion in section 4.6 it was argued that the consequent stages of the ΣΔ ADC are likely to consume less than the corresponding analog stages in the conventional channel. In addition, the discussion in section 3.3.2 demonstrated that the accuracy requirements on the analog sections in the conventional channel are strict as compared to those on the consequent stages of the ΣΔ ADC. Hence, it may seem that the power consumption of the multi-channel ΣΔ ADC could even be lower than that of the analog part of the conventional channel. *However, up to now only noise and distortion have been considered. The true challenge on multi-channel A/D conversion lies in sample rate related problems.*

As discussed above, the sample rate of the multi-channel ADC may become very high. The bandwidth of the sub-circuits of the multi-channel ΣΔ ADC must be related to this high sample rate while the bandwidth of the analog circuits in the conventional channel is only related to the -much smaller- signal bandwidth. Hence, *for bandwidth reasons the current consumption of the multi-channel ΣΔ ADC is likely to be much higher than that of the input stage of the conventional channel.*

Digital part: With m_1 being the over-sampling factor for the conventional ΣΔ and m_2 the over-sampling factor for the multi-channel architecture, the following worst-case power increase is to be expected (eq. 4.19 but neglecting the small influence of the logarithmic function):

$$\frac{I_{\text{digital}}(\text{digital CCh})}{I_{\text{digital}}(\text{analog CCh})} = \frac{m_2 p \left[\log\left(DR_i\, q^2/p\right)\right]^3}{m_1 \left[\log SINAD_o\right]^3} \tag{5.2}$$

The higher resolution demand only puts a relatively small penalty on the power consumption in digital because of the logarithmic relation. On the contrary, the increase in clock rate has a major effect. So, also from a digital perspective multi-channel conversion is not necessarily an attractive option yet. Of course, scaling laws are beneficial here.

The discussion is illustrated with an example. A Bluetooth application is considered because -when compared to UMTS, GSM, a.o.- this is a relatively "easy" application. Therefore, it may be a candidate for digitization.

Example: Bluetooth

A Bluetooth channel is about 1MHz wide, the entire band spans 78MHz (hence $p = 78$). Digital demodulation requires a resolution of about 18dB. The amplitude range of one channel is 50dB. The maximum interferer amplitude is as large as the maximum wanted signal, i.e. $q = 1$.

Taking sufficient margin, we compare:

- *an ADC for the full-analog conditioning channel, with SINAD = 40dB (i.e. ≫ 18dB)*[3] *realized using $m_1 = 16$ and $L = 3$*

[3] Comparison to an ADC with only 18dB of *SINAD* is unrealistic. First of all, because that requires a lot of variable gain and gain control and therefore becomes un-practical. Secondly, it is unlikely that a ΣΔ ADC would be used for such low requirements.

- *an ADC for the full-digital conditioning channel, with SINAD=70dB (i.e. > (18 + 50)dB) realized using m_2=40 and L=4 to 5*

Then, for noise and distortion reasons, eq. 5.1 predicts:

$$\frac{I_{\text{ADC}}(\text{digital CCh})}{I_{\text{ADC}}(\text{conventional})} = 1\left(\frac{10^7}{10^4}\right)^{7/6} 10^{4/3} \cong 1e5$$

Still, it is expected that the calculated power increase in the ADC corresponds to what is gained by omitting the analog conditioning circuits. Instead, the dominant power penalty is due to higher bandwidth requirement. In the analog channel the ADC is sampled at 32MHz, in the digital channel at 6.24GHz. This is far beyond what is feasible at present (see table 3.2).

The current consumption of the decimation filter increases according to eq. 5.2:

$$\frac{I_{\text{digital}}(\text{digital CCh})}{I_{\text{digital}}(\text{analog CCh})} = \frac{40 \cdot 78\left[\log\left(10^7/78\right)\right]^3}{16\left[\log 10^4\right]^3} \cong 400$$

The input sample rate of 6.24Gs/s is hardly feasible in present-day CMOS technology and at least not common for DSPs.

This example shows that, even for a standard as "easy" as Bluetooth, a full-digital signal conditioning channel does not seem feasible in view of state-of-the-art ΣΔ ADC performance and speed of DSP.

Analog reference circuits

Although the power consumption in the peripheral circuits (such as time, voltage or current references) is not systematically analyzed in this book it is worth putting a remark here. These circuits generate a reference for the ΣΔ ADC (i.e. for the feedback DAC). Their accuracy must be as good as what is required of the ΣΔ ADC as a whole. Clearly, the multi-channel ΣΔ ADC requires high-precision and, by consequence, power-hungry references. Especially, the clock-generation is extremely challenging: not only does it need to have low jitter, it must also provide a high frequency.

From the discussion and the example, it becomes clear that full-digital signal conditioning puts impressive requirements on the ADC. In view of state-of-the-art ΣΔ ADC performance and DSP capability, full-digital signal conditioning is not yet feasible for systems operating over a large frequency band with a large dynamic range. Hence, this is the case for most wireless systems.

An exception is presented in chapter 8. There an AM-radio receiver with multi-channel A/D conversion is discussed. AM radio is a narrow-band system. As such, the sample rate of the multi-channel ADC does not become excessive. Instead, it enables a power-effective alternative to a full-analog channel and provides a lot of flexibility.

5.3 Conclusions

Full-digital signal conditioning with multi-channel A/D conversion puts impressive requirements on the $\Sigma\Delta$ ADC. The sample rate of the ADC and the digital circuits increases dramatically, possibly exceeding state-of-the-art sample rates for a $\Sigma\Delta$ ADC and common DSP speed.

Straightforward digitization results in a major increase in the power consumption of the ADC and of the overall channel.

Full-digital signal conditioning is not yet feasible for systems operating over a large frequency band with a large dynamic range. This is the case for most wireless systems.

The discussion in this chapter is illustrated with example implementations in chapter 8.

Chapter 6

Conditioning ΣΔ ADCs

Because of the over-sampling and the continuous-time loop filter, a certain level of interferers can be applied to the ΣΔ ADC without corrupting the resolution in the bandwidth of the wanted channel. This characteristic of the continuous-time ΣΔ ADC is exploited in a third architecture for the conditioning channel. The ΣΔ ADC is used without preceding analog filters or VGA. Instead, its inherent interferer immunity is exploited or even enhanced. This is achieved by moving some signal conditioning into the ΣΔ loop. The ΣΔ ADC, used in this manner, is referred to as a "conditioning ΣΔ ADC".

First, the concept of a "conditioning ΣΔ ADC" is introduced along with an overview of architectures. Next, a universal model of a generic ΣΔ ADC is described. This model is used to analyze the major limitations on the allowable interferer level for the ΣΔ ADC. Then, various topologies for the conditioning ΣΔ ADC are presented. These topologies differ in the amount of signal conditioning that is integrated into the ΣΔ loop and in the power/performance relation of the overall channel. They are compared along the lines developed in chapter 5.

6.1 Generic conditioning ΣΔ ADC

In chapter 5 channels with full-analog and full-digital signal conditioning have been presented (fig. 6.1.a and b respectively). Here, the signal conditioning is integrated into the ΣΔ ADC resulting in a "conditioning ΣΔ ADC" without preceding analog filters or variable gain stages (fig. 6.1.c). First, the basic operation of the conditioning ΣΔ ADC is clarified. Next, a universal model of a ΣΔ ADC is introduced. This model is used for a generic analysis of the interferer immunity of continuous-time ΣΔ ADCs. At the end of this section, a basic power/performance analysis is presented.

6.1.1 Concept of operation

The concept of implementing some signal conditioning into the ADC, is developed in three steps. The solutions presented in sections 6.2 and 6.3 exploit characteristics of

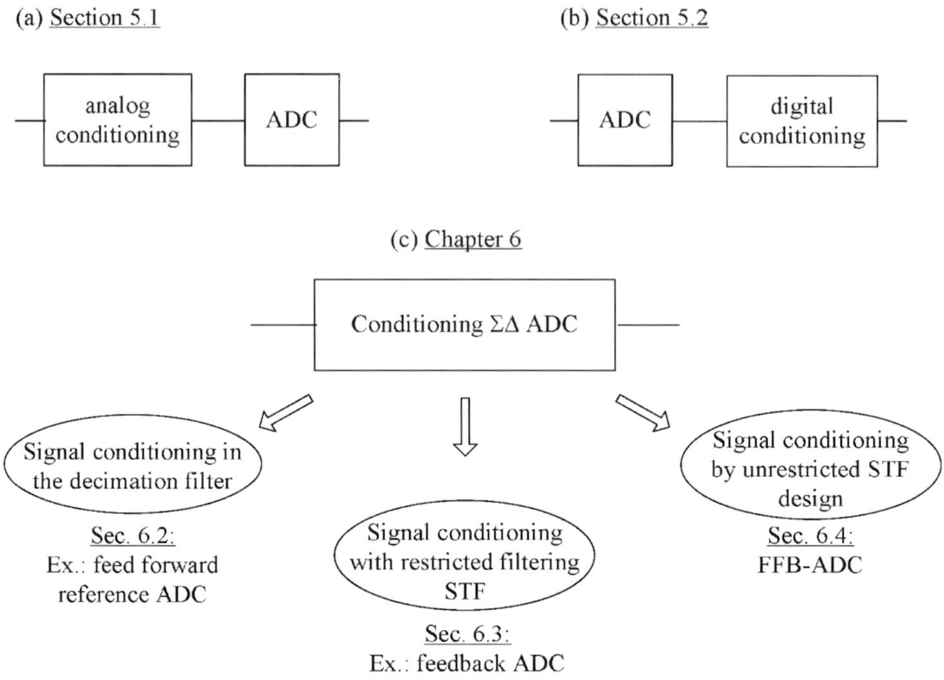

Figure 6.1: *Full-analog (a), full-digital (b) and mixed-signal (c) signal conditioning*

known ΣΔ topologies. The solution presented in section 6.4 realizes the signal-conditioning by explicit design. The consecutive solutions build on each other's properties. They are introduced only briefly at present but are detailed later in this chapter.

Signal conditioning in the decimation filter

Like in the full-digital architecture of section 5.2 both the wanted and the interferer channels are applied to the ΣΔ ADC, i.e. no analog pre-filtering or VGA is applied. A key difference compared to the full-digital architecture is the fact that the conversion bandwidth of the ADC is only as large as the bandwidth of the wanted signal. The operation of the conditioning ΣΔ ADC is intuitively illustrated in fig. 6.2 for the same composite input signal as used in chapter 5. The basic characteristic enabling this operation, is the fact that a ΣΔ ADC uses over-sampling. On top of that, in case of a continuous-time implementation, the loop filter acts as an anti-alias filter. This results in some *immunity of the ADC to interferers; i.e. interferers -below a set amplitude limit- can be applied without affecting the resolution in the bandwidth of the wanted channel.* The limit on the immunity is analyzed in a more quantitative way in section 6.1.3.

Using the ΣΔ ADC in this way, the interferer channels appear in the noise-shaped part of the output spectrum. The ADC provides frequency selective resolution; i.e. the

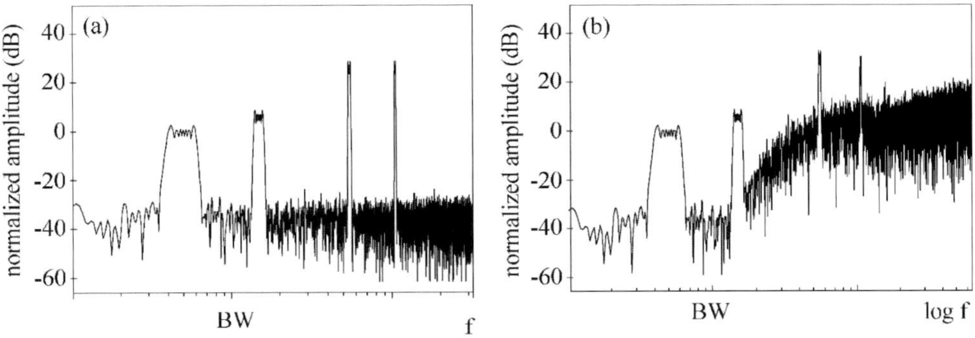

Figure 6.2: *Input spectrum (a) and output spectrum (b) of a conditioning ΣΔ ADC*

wanted channel is digitized accurately while interferer channels are de-correlated by the shaped quantization noise. Next, the decimation filter attenuates the interferers, along with the quantization noise. This is illustrated in fig. 6.3.a: the interferers pass through the ADC[1] but are attenuated by $H_{dec}[z]$. The latter also includes the digital VGA.

Note that, *while in the full-digital architecture of fig. 5.3 a dedicated digital channel filter $H_{ch}[z]$ is needed, here, this functionality is inherent to the decimation filer $H_{dec}[z]$.* At the input of the decimation filter, the "unwanted energy" now includes both the quantization noise and the interferers. Still, the total "unwanted energy" is the same as in the situation without interferers. It equals the energy in the bitstream, which is a constant, minus that in the wanted signal. Thus, it is independent of the interferers. On the contrary, the *distribution* of the "unwanted energy" can be different. Still, in practice, this should hardly affect the attenuation requirements because the decimation filter must anyway be designed to suppress worst-case tones. Similarly, the word-length scaling that happens in a decimation filter, simultaneously fulfills the role of digital VGA.

Finally, notice the following:

- while the signal-conditioning happens in the decimation filter, still, this solution is classified as a conditioning ΣΔ ADC. Since the decimation filter is indispensable for ΣΔ A/D conversion, it can be considered part of the "broader ADC";
- the signal-conditioning in the decimation filter is only possible because of the interferer immunity of the continuous-time ΣΔ ADC (see section 6.1.3).

In section 6.2, the channel with signal-conditioning in the decimation filter, is analyzed for the example of a feed forward ADC, as a reference[2].

While having economized on the channel filter and VGA, some analog circuits in

[1] Note that, in this example, the interferers are amplified throughout the ΣΔ ADC; i.e. the transfer of the interferers of amplitude $q\,\hat{v}_{\text{MAX}}$ shows a weak, positive slope. This is due to the frequency dependence of the signal transfer function (STF) of the ΣΔ ADC. The inclination and direction of the slope should be considered as an example since they depend on the STF of the ΣΔ ADC and on the frequency of the interferers.

[2] In fact, any continuous-time ΣΔ ADC can be used in this way because this conditioning channel is enabled by the over-sampling that is inherent to ΣΔ modulation and by the anti-alias suppression of the loop filter

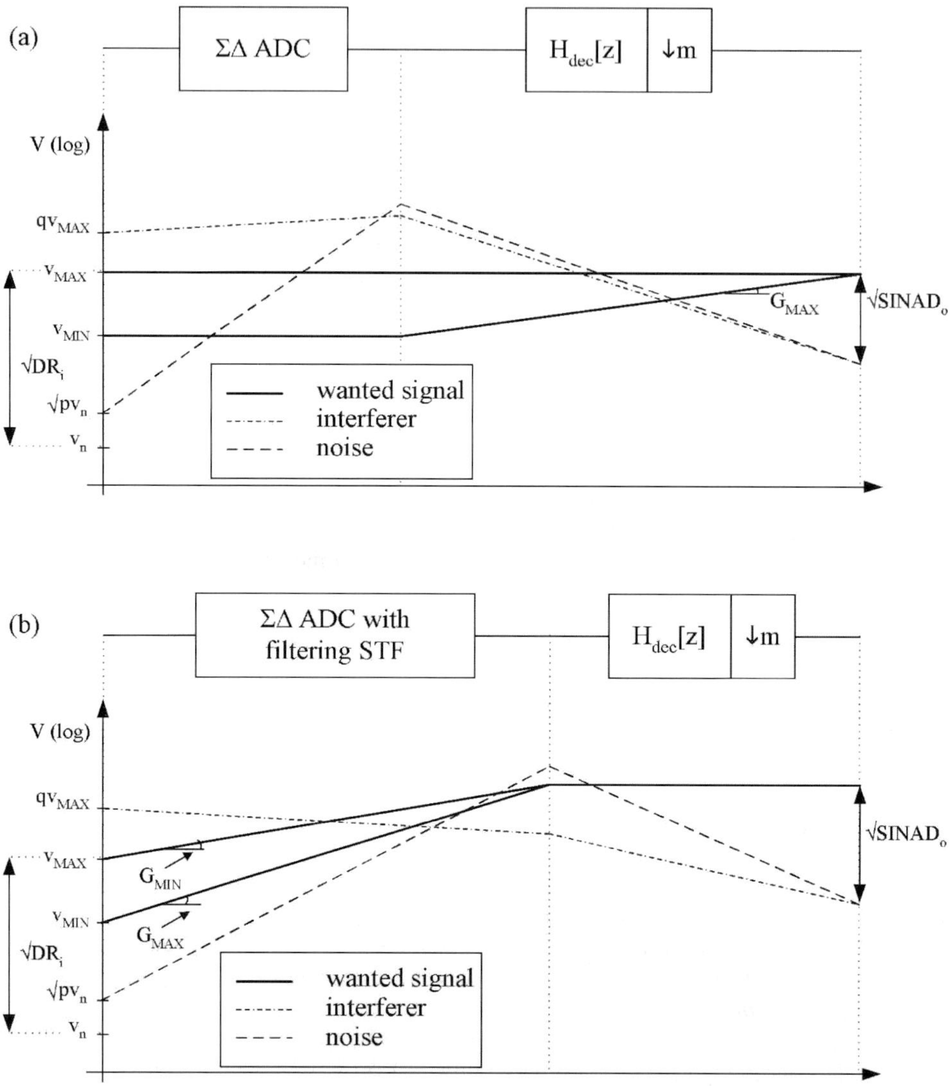

Figure 6.3: *Conditioning ΣΔ ADCs with signal-conditioning in the decimation filter (section 6.2) (a) or using a filtering STF (sections 6.3 and 6.4) (b)*

this channel architecture remain very difficult. This is alleviated in the following solutions using signal-conditioning in the ADC itself (next to conditioning in the decimation filter).

Signal conditioning with restricted filtering STF

Next to its inherent interferer immunity, a second characteristic of a ΣΔ ADC can be exploited; i.e. *the signal transfer function (STF) of a ΣΔ ADC is not flat, instead, it shows frequency selectivity.* If the frequency selectivity is favorable -i.e. $|STF| < 1$ for interferers channels while $|STF| = 1$ for wanted channels- this property can be used for implementing signal-conditioning into the ADC. In section 6.3, it is shown that ΣΔ ADCs based on a loop filter with nested feedback, fulfill this requirement. Then, the STF provides filtering[3] from the input to the output of the ADC. In addition, VGA functionality is easily added either by varying the input resistance or the full-scale level of the feedback DAC. Note that the achieved filtering is restricted in the sense that the STF and the noise transfer function (NTF) cannot be optimized independently.

The signal-conditioning is further illustrated in fig. 6.3.b. The interferers at an input level $q\ \hat{v}_{\mathrm{MAX}}$ are attenuated throughout the ΣΔ ADC. On the contrary, the wanted signal is amplified with a gain ranging from G_{MIN} to G_{MAX} depending on the signal strength. Hence, filtering and VGA are integrated into the ΣΔ ADC and the same functionality is realized as in the conventional cascade of filter, VGA and ΣΔ ADC in fig. 5.2. It can be concluded that, if the STF of the ΣΔ ADC is favorable, a true "conditioning ΣΔ ADC" can be implemented.

Signal conditioning by unrestricted STF design

While, in the above solution a given STF of a known ΣΔ topology, is *exploited* for signal-conditioning, the STF of the ADC may as well be *designed for* signal-conditioning. Dedicated design of the STF, allows for *optimization according to the desired filtering characteristic. Contrary to the above solution with restricted filtering, this is possible without compromising on the noise shaping. In addition, a more power-efficient ΣΔ topology can then be used.* A conditioning ΣΔ ADC with this property is presented and analyzed in section 6.4. It is referred to as a "filtering-feedback ΣΔ ADC" or FFB-ADC. The diagram of fig. 6.3.b. remains valid for this ADC.

All solutions with a conditioning ΣΔ ADC rely on the inherent interferer immunity of a continuous-time ΣΔ implementation. In fact, it is shown in section 6.3 and 6.4 that the solutions with signal-conditioning integrated into the ADC are enabled by an *improved interferer immunity* of the design.

[3] "Filtering" is used as a synonym for "frequency selectivity". In this book, this term does not imply *full* attenuation of interferers.

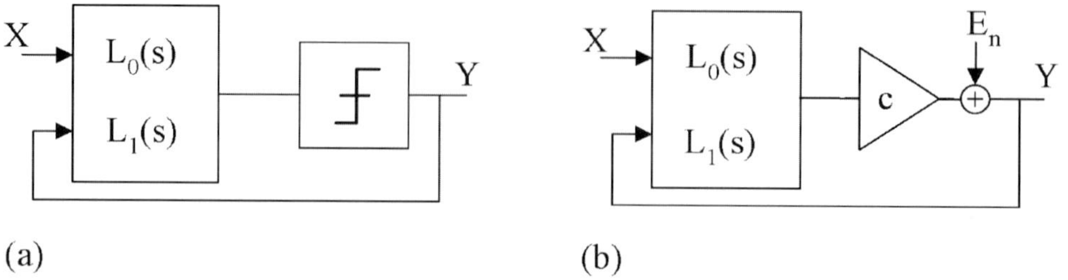

Figure 6.4: *Universal model of a 1-bit ΣΔ modulator (a) and linearized equivalent (b)*

6.1.2 Universal model of a ΣΔ modulator

A universal model of a single-bit ΣΔ modulator is depicted in fig. 6.4.a [73], [25]. This model captures all ΣΔ topologies that are discussed further on and therefore enables a generalized discussion[4]. It consists of a linear block comprising the filter functions L_0 and L_1 and a non-linear block consisting of the 1-bit quantizer. The DAC in the feedback path is implicit. L_0 and L_1 are filters acting on the input signal and on the feedback signal respectively. A simple linearized model is shown in fig. 6.4.b where the quantizer is replaced by a linearized gain c and an additive noise source E_n. A calculation of c and E_n is a.o. documented in [25], section 4.2.1. The signal transfer function of the ΣΔ ADC can be calculated in the s-domain:

$$\frac{Y(s)}{X(s)} = \frac{c\,L_0(s)}{1-c\,L_1(s)} \tag{6.1}$$

The noise transfer function (NTF) equals:

$$\frac{Y(s)}{E_n(s)} = \frac{1}{1-c\,L_1(s)} \tag{6.2}$$

Numerous publications treat the optimization of L_1 to improve the NTF while maintaining stability [74], [75], a.o. On the contrary, relatively little attention has been paid to the STF of the ΣΔ modulator. Publications that do treat STF-design especially target a flat pass-band response for audio applications [76], [77]. Here, the STF-design is exploited to improve the immunity of ΣΔ ADCs to interferers outside the ADC pass-band.

The model of fig. 6.4 is introduced in view of a generic analysis of various ΣΔ topologies introduced further on. All those topologies realize the same L_1 and thus the same NTF (given a fixed order for the loop filter). The loop filter L_0 is designed to optimize the STF in view of interferer immunity.

6.1.3 Interferer immunity

The over-sampling principle used in ΣΔ ADCs is the basic reason for their robustness to interferers outside of the conversion bandwidth. As the amplitude of the interferers

[4]Here, this model is meant for describing only the ADC. More generalized, it can be used to include the analog conditioning in front of the ADC as well. Then, this functionality would be part of the $L_0(s)$-function.

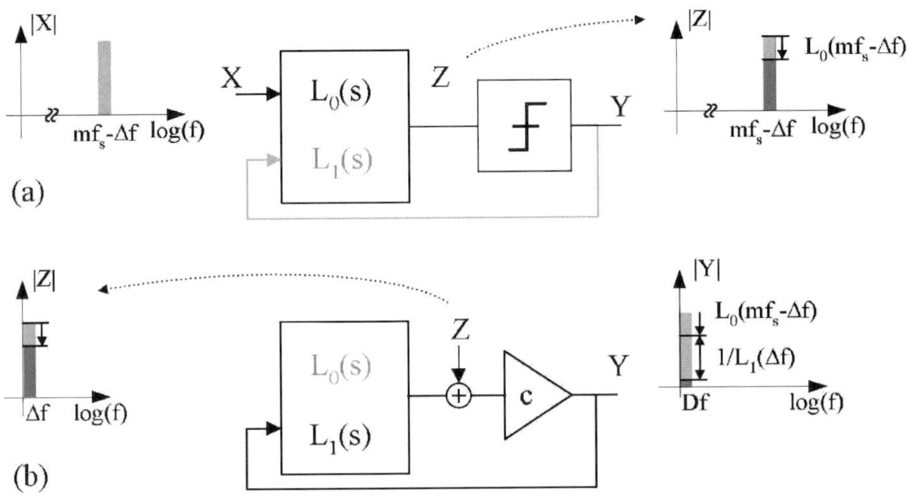

Figure 6.5: *ΣΔ ADC (a) and linearized model (b) for analyzing the anti-aliasing suppression*

exceeds a set limit, though, the interferers do cause distortion, spurious responses and an increase of noise in the wanted channel. This limit is due to various effects and can be frequency dependent. These effects are analyzed and quantified next.

Aliasing

Compared to ADCs sampled at the Nyquist rate any ΣΔ ADC benefits from reduced anti-aliasing requirements because of the over-sampling:

- the number of frequency bands that fold back into the conversion band is less than in case of sampling at the Nyquist frequency.

Only continuous-time implementations benefit from an additional characteristic:

- the interferers near the clock frequency are suppressed by the loop filter before being sampled in the quantizer.

The latter is not valid for discrete-time implementations: by definition, the transition to the discrete-time domain -i.e. the sampling action- occurs already at the input of the discrete-time ADC. The aliasing that occurs at this point is irreversible: the in-band spurious cannot be removed by the loop filter.

A quantitative analysis of the alias-suppression is conducted based on fig. 6.5. Suppose an interferer at a frequency $mf_s - \Delta f$ near the sample frequency is applied to the continuous-time ΣΔ ADC. Typically, the unity-gain frequency of L_1 is at $1/6^{th}$ of the sample rate [25] for stability reasons. This implies that the feedback is not effective anymore at the interferer frequency and the feedback path in fig. 6.5.a can be neglected.

The interferer is attenuated by loop filter L_0 and consecutively sampled in the quantizer. This introduces an alias Z at Δf. Fortunately, this alias Z is suppressed towards the output because of the high preceding gain at frequency Δf. The (approximate) transfer of the alias $Z(\Delta f)$ towards the output $Y(\Delta f)$ of the ADC can be calculated from the simple linearized model of fig. 6.5.b. The overall suppression then equals:

$$\left|\frac{Y(\Delta f)}{X(mf_s - \Delta f)}\right| = \left|\frac{c\,L_0(mf_s - \Delta f)}{1 - c\,L_1(\Delta f)}\right| \tag{6.3}$$

$$\approx \left|\frac{L_0(mf_s - \Delta f)}{L_1(\Delta f)}\right| \tag{6.4}$$

As mentioned, L_1 is kept constant in the remainder while various L_0-functions are considered for improving a.o. the alias-suppression.

Stable input range

In section 3.2.3 the stable input level was defined implicitly, assuming only a signal within the conversion bandwidth of the ΣΔ ADC is applied. *Here, this definition is extended to inputs outside the conversion bandwidth. For these frequencies, the stable input range is defined as the input level for which the maximum modulation depth of the output is reached.* As such, this definition is an extrapolation of that in section 3.2.3, where it was noted that, when applying an input amplitude as large as the stable input level, then the modulation depth of the output is maximal.

As motivated in section 3.2.3 the maximum modulation depth of a well-designed single-bit ΣΔ modulator is about 70%. This corresponds to a value of -3 dB compared to digital full-scale. In the analog domain and assuming a DAC gain of 1, this is a value of $0.7\,v_{DAC}$. v_{DAC} indicates the reference voltage of the DAC. The corresponding input level, i.e. the stable input range, can then be derived by means of the STF of eq. 6.1:

$$\text{Stable input range}(j\omega) = \frac{0.7\,v_{DAC}}{|STF(j\omega)|} \tag{6.5}$$

Although the maximum output modulation depth is constant over frequency[5], the stable input range is not. Its frequency dependence is inversely proportional to that of the STF. Therefore, it can be improved by proper design of L_0. (As mentioned, L_1 is kept constant to realize optimal noise shaping.)

Notice that the above limit is derived from a linearized model of the ΣΔ ADC. As such, the results are approximate and only valid within the small-signal operation regime. Verification of this limit from the linear analysis with measurements in chapter 9 yields good correspondence. Hence, the model is adequate for determining the transition to instability and large-signal behavior.

[5] In fact, for higher signal frequencies the maximum modulation depth of a bit-stream code increases slightly. This effect is marginal and is disregarded in the remainder.

Spurious responses

The phenomenon of "spurious responses" in a ΣΔ modulator is due to the fact that its quantization noise is not random. Instead, patterns -correlated to the input signal- may appear at the output of the ADC in some, often not obvious way. In the frequency domain, this translates into tones.

In this book, the term "spurious responses" is used in a limited sense. Here, we focus on the tones that are generated when an interferer is applied near even sub-harmonics of the sample frequency, e.g. $mf_s/2$, $mf_s/4$, *etc. In practice, the analysis can be further narrowed to inputs applied near* $mf_s/2$ *since these are most notorious.*

First, available literature on the analysis of spurious tones is referenced. Next, the mechanism causing spurious responses in a ΣΔ ADC is briefly recapitulated and some techniques to reduce spurious tones are listed. Finally, it is noticed that, in practice, many spurious responses may just as well be due to deficiencies in the analog circuits or the layout rather than being caused by the tonal behavior of ΣΔ encoding.

Analysis: Spurious responses are well-studied for low-order modulators and DC-inputs [78], [79]. For higher-order modulators the results become approximate and, again, assume DC-inputs [80]. *To the author's knowledge, no analytical studies on spurious responses of high-order modulators due to high-frequency inputs have been published yet. The discussion in this section remains qualitative as well. It focuses on pragmatic guidelines to reduce spurious responses caused by high-frequency interferer inputs.*

Spurious responses for low-frequency inputs: This phenomenon is intuitively understood from the theory on asynchronous ΣΔ modulators in [81] and considering the effect of sampling next.

First, when applying a low-frequency input to an *asynchronous* ΣΔ modulator, the following occurs:

- apart from the wanted output, also odd harmonics are generated;
- the input causes FM-modulation of the idling frequency of the ΣΔ modulator. This results in a Bessel-spectrum around $mf_s/2$ and around harmonics of this frequency.

Second, considering the sampling in a *synchronous* ΣΔ ADC [80], [82]:

- aliasing of the spurious tones occurs.

For this example of a low-frequency input, the latter means that a.o. the Bessel components around the second harmonic of the idling frequency appear at baseband. Fortunately, all the described tone-generating mechanisms occur in the quantizer. By consequence, the tones appearing at baseband are suppressed by the noise shaping of the loop.

Notice that, for a small input signal, the amplitude of the spurious tones varies monotonically with the input amplitude. This is obvious for the odd harmonics but is also true for the Bessel-components. In fact, under this condition, especially the 1st-order Bessel component is generated around $mf_s/2$. Moreover, the amplitude of this component is more or less linear with the input amplitude. For larger inputs, more and more Bessel components become important. In addition, their amplitude varies very irregularly as a function of the input amplitude.

Spurious responses for high-frequency inputs: In the present context of interferer immunity of a ΣΔ ADC, we focus on inputs applied near $mf_s/2$. This input frequency is above the unity-gain frequency of the loop. As such, the signal is simply attenuated by L_0 before being applied to the quantizer. In the quantizer, the non-linear, tone-generating mechanisms, occur, similarly as described above[6]. However, on the frequency axis, the picture looks very different:

- the odd harmonics of the input signal are at high frequencies. Moreover, they are maximally separated from the sample frequency mf_s (or harmonics thereof). As such, aliasing will not cause components in the baseband;
- FM-modulation of the idling frequency directly results in a Bessel component appearing at baseband. In case of a small input or strong attenuation of the input signal by L_0, this is the most important component and its amplitude is almost linear with the input amplitude;
- the Bessel components generated at high frequencies fold back due to aliasing. This mechanism becomes important for larger input amplitudes.

Again, all these mechanisms occur in the quantizer. As such, the tones at low-frequencies, are counteracted by the noise shaping.

In this book, we look for the allowable amplitude of inputs around $mf_s/2$, such that the amplitude of the first Bessel component does not deteriorate the *DR* in the wanted channel. As mentioned, an accurate, quantitative analysis of this problem is not described in literature, nor do we aim for it here. Instead, this limit is evaluated experimentally for the conditioning ADCs presented in chapter 9. In addition, we present techniques to reduce spurious responses for these inputs.

Reduction of spurious responses: Common techniques for reducing spurious responses aim at de-correlation between the input signal and quantization noise. This can be achieved by:

- stronger noise shaping; i.e. increasing the order of L_1;
- adding “dither”.

The latter technique assumes that the frequency of the dithering signal is larger than that of the input signal causing the tones. As such, dithering may not be effective for the high-frequency input that is considered here. In addition, a lot of dithering energy may be required to lower the tones [83]. Part of the dithering signal appears in the wanted channel and, consequently, affects the *DR*.

[6]Since the input signal is applied beyond the unity-gain bandwidth of the ΣΔ modulator, this is an extrapolation of the theory for low-frequency inputs. Consequently, this should be considered as a first-order approximation only.

Alternative to the above, the high-frequency interferer can be attenuated before being sampled in the quantizer. Possible implementations are:

- a simple (notch) filter with stop-band around $mf_s/2$ preceding the ADC;
- design of L_0 for additional suppression near $mf_s/2$.

Although the first solution is based on preceding, analog signal conditioning the filter can be much easier than the filter in the full-analog conditioning channel of fig. 5.2. In this book, the second means, i.e. L_0-design, is used.

Spurious tones due to analog deficiencies: In practice, spurious tones appearing in the bandwidth of interest, may as well be due to a very different effect. As such, they may be caused by:

- parasitic mixing of the input with multiples of $mf_s/2$ (for example due to cross-talk from divided clocks on the DAC reference, bias lines or supply rails) or with the high-frequency quantization noise in the output spectrum (for example present on the substrate, supply or bias lines);
- IM_3-distortion of the input and the feedback signal may occur in the analog blocks.

These problems are common to all analog circuits and can be prevented by linear design of the input stage, input and feedback paths and by careful layout.

The above discussion does not enable a quantitative prediction of the tones. Still, in practice the problem of spurious responses is manageable when taking into account the mentioned guidelines. This pragmatic approach is in fact similar to the "acceptance" of high-order ΣΔ designs. Although a solid proof of stability lacks, still, numerous high-order modulators have been designed successfully or even proven in mass-production [77], [38], [2], [37], etc. In chapter 8, measurements on spurious responses are included for various designs.

Intermodulation distortion

As interferers are applied to the ΣΔ ADC care must be taken that intermodulation distortion is sufficiently low. Intermodulation of interferers and/or shaped quantization noise could corrupt the resolution in the conversion bandwidth of the ΣΔ ADC. Therefore, the IM_3-performance of the ΣΔ ADC yields another limit on the maximum allowable input level for interferers. This limit can be calculated from eq 4.12 assuming the transconductance of the input stage remains the dominant cause of distortion over a large frequency range.

Obviously, this particular limitation cannot be relieved by the choice of L_0.

6.1.4 Power/performance analysis

The power/performance analysis of the channel is strongly dependent on the amount of signal conditioning that happens in the ΣΔ ADC. Therefore, the discussion here is limited to the basic requirements on the ADC and is detailed later on.

Noise and distortion

The input-referred noise and distortion requirements on the conditioning ΣΔ ADC are the same as those on the input stage of the full-analog channel and those on the ΣΔ ADC in the full-digital channel. Hence, the specifications in the last column of table 5.1 apply for the conditioning ΣΔ ADC (except for $p = 1$ since, here, the conversion bandwidth corresponds to the wanted channel only). In an analogy to eq. 5.1, the quiescent current required to meet the noise and distortion requirements follows:

$$\frac{I_{\mathrm{ADC}}(\Sigma\Delta \text{ based CCh})}{I_{\mathrm{ADC}}(\text{analog CCh})} = F\, q \left(\frac{DR_i}{SINAD_o}\right)^{7/6} G_{\mathrm{MIN}}^{4/3} \tag{6.6}$$

The factor F is an implementation factor used to generalize the expression. All the conditioning ADCs that are discussed in this chapter, target the same performance and obey the same power/performance relation. Still, their absolute power consumption is different. This difference is captured in the factor F. The value of F is determined by the ΣΔ topology. By definition $F = 1$ for the feed forward reference ADC, discussed in section 6.2; i.e. that ADC is taken as a reference. The ADCs of sections 6.3 and 6.4 have $F \neq 1$.

Sample rate

The advantage of this third architecture compared to the full-digital architecture, is in its lower sample rate mf_s:

- the Nyquist sample rate f_s remains as low as in the full-analog channel;
- the over-sample ratio m must be increased though to keep quantization noise low enough in view of the high DR specification.

As a result, for the reference ADC, the sample rate can be lower by a factor p, when compared to the full-digital architecture. For the conditioning ΣΔ ADCs with a filtering STF and VGA, the over-sampling can be reduced further. This is shown in sections 6.3 and 6.4.

Circuit bandwidth

Depending on the application, the circuit bandwidth is set either by the sample rate of the ΣΔ ADC or by the bandwidth of the input signal. The latter may be quite large since interferer channels can be present over a wide band. This is illustrated in table 6.1 listing the bandwidth of a wanted channel[7] and of the entire band for a few wireless communication systems. For most examples, it is likely that the sample rate of the ΣΔ ADC is

[7] In fact, the channel spacing is listed instead of the channel bandwidth. The channel bandwidth is only determined by the modulation. The channel spacing also takes into account guard bands in between neighboring channels.

Table 6.1: *Channel bandwidth and operation band for a few wireless communication systems*

System	Channel	Band	mf_s and ref.
GSM	200kHz	25MHz	26MHz in [84]
UMTS	5MHz	60MHz	153.6MHz in [84]
Bluetooth	1MHz	78MHz	64MHz in [32]
IEEE802.11a	20MHz	100MHz	160MHz in [85]

in the same order of magnitude as the operation band. In those cases, the wide range of interferer channels doesn't put a significant penalty on the circuit bandwidth.

In the following sections various implementations of a conditioning ΣΔ ADC are presented. The signal conditioning ranges from "exploiting" the properties of known ΣΔ topologies in sections 6.2 and 6.3 to explicitly moving an analog filter into the ΣΔ loop in section 6.4.

6.2 Signal conditioning in the decimation filter

This conditioning channel is analyzed for the example of the feed forward ADC depicted in fig. 6.6, as a reference case. Generalized, an Nth-order loop filter is considered. Here, $N=3$ as an example. Referring to the model of fig. 6.4 L_0 and L_1 can be calculated:

$$L_0(s) = -L_1(s) \tag{6.7}$$

$$= a_1H_1(s) + a_2H_1(s)H_2(s) + a_3H_1(s)H_2(s)H_3(s) \tag{6.8}$$

In this topology, only the error signal is processed in the loop filter. The first filter stage can have high gain and, by consequence, the following stages are not critical. This results in a very low-power implementation. *Implementing the highest possible gain in the input stage, is a general advice for a low-power design strategy.*

This ΣΔ ADC can be used in the conditioning channel of fig. 6.3.a. In the following, the tolerance of this topology to interferers is emphasized rather than its performance for wanted channels. This topology is referred to as the "reference conditioning ΣΔ ADC", or, even shorter, as the "reference ADC".

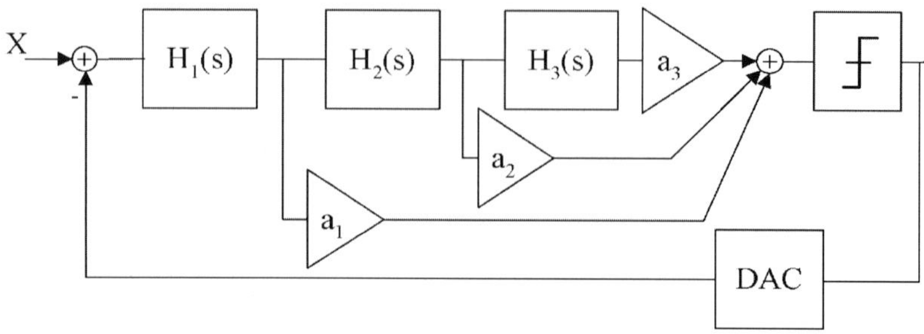

Figure 6.6: *Reference ADC with a 3rd-order feed forward loop filter*

6.2.1 Interferer immunity

In section 6.1.3, the interferer immunity was analyzed for the universal ΣΔ modulator of fig. 6.4. Here, those results are applied to the reference ADC.

Aliasing

Filling out eq. 6.3, the filter L_0 only has first-order roll-off for inputs near the clock frequency due to the feed forward paths. Typically, the unity-gain frequency of the loop filter is at $1/6^{th}$ of the sample rate [25] for stability reasons. Hence, the associated suppression is rather small ($\sim -15dB$). The consequent aliasing that occurs in the quantizer is suppressed towards the output by the Nth-order noise-shaping of the loop. Hence, the overall alias-suppression function is of order $N+1$.

Stable input range

The stable input range (eq. 6.5) of this topology can be derived from the STF. Based on eq. 6.8 and eq. 6.1 the loop gain (equaling $c\,L_1$ for the model of fig. 6.4) and the magnitude of the STF of this ΣΔ ADC are depicted in fig. 6.7.a and b. The exact curve strongly depends on the actual parameters of the loop. Hence, these graphs should merely be considered as an example showing some general characteristics. Below the unity gain frequency f_{ug} of the loop the STF is constant except for the overshoot of several decibels adjacent to the conversion bandwidth. This is due to the fast phase shift of the open loop gain around f_a (f_a indicates the transition from a high-order to a first-order slope in the loop gain) and translates into a reduced stable input range for adjacent interferers. Far-off interferers -beyond the unity-gain of the loop- are suppressed with a first-order slope corresponding to the first-order open loop transfer.

Spurious responses

The comments of section 6.1.3 apply.

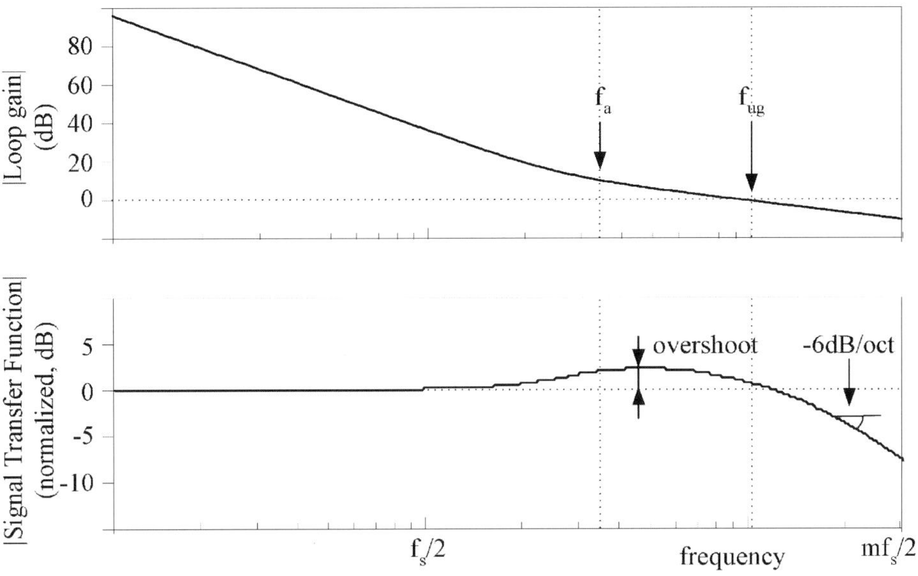

Figure 6.7: *Magnitude of the loop gain and the STF of the reference ADC of fig. 6.6*

Intermodulation distortion

Again, the discussion in section 6.1.3 is valid since the dominant non-linearity of this topology is normally due to the transconductance of the input stage.

The above limitations on the allowable level of interferers are summarized in fig. 6.8 for the 3rd-order topology of fig. 6.6. The four limitations are indicated by a dashed line. The solid curve indicates the dominant limitation over frequency. Of course, the relative position of the various curves strongly depends on the actual implementation and this graph should be considered as an illustration. Still, for most frequencies, the allowable interferer level of the reference ADC is often limited by the maximum stable input.

6.2.2 The conditioning channel

As analyzed above, the reference ADC is robust to interferers outside the conversion bandwidth of the ΣΔ ADC. By consequence, the conversion bandwidth is designed to accommodate the wanted channels only and a bandwidth efficient solution results. Resolution-wise, though, this efficiency cannot be found. Just as in the multi-channel ΣΔ ADC the resolution requirements are set by the interferers instead of the wanted signal:

- *DR* and IM_3 are related to the entire input signal including the interferers;
- the stable input range of the ΣΔ ADC is dimensioned to the largest interferer while the wanted signal can be smaller.

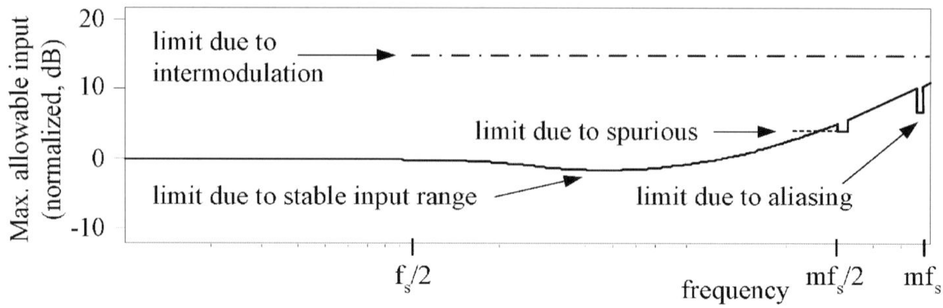

Figure 6.8: *Allowable input level for the reference ADC of fig. 6.6*

This is illustrated with an example next.

Example:
The input signal of fig. 6.9.a consists of a weak wanted channel and strong interferer channels. The stable input range of this ADC is dimensioned to accommodate the interferers. Therefore, the "upper part" of the available DR is not used efficiently. In fig. 6.9.b the wanted channel is strong. It is converted with a very large SINAD while essentially much less resolution is required for further processing.

These observations of the "wasted" *SINAD* or *DR* indicate that the ΣΔ channel of fig. 6.3.a does not necessarily yield the best channel architecture in terms of power efficiency. Improved solutions, with signal conditioning in the ΣΔ ADC, are presented in the following sections.

6.2.3 Power/performance analysis

To a large extent, the generic analysis of section 6.1.4 applies. Some parameters are detailed here.

The conditioning channel

The power consumption of the analog and the digital part of the conditioning channel are -separately- compared to that of the full-analog and the full-digital channel in chapter 5.

Analog part: The generic discussion on the power consumption of a conditioning ΣΔ ADC (see page 77) fully applies. For noise and distortion reasons, the input stage consumes as much as any of the other channels. The bandwidth requirements are much more realistic than those for the full-digital channel, because of the lower sample rate. Likely, the circuit bandwidth must be dimensioned to the bandwidth of the applied interferers. In that case, this specification -and the consequent power consumption- is comparably difficult as that of the input stage of the analog channel.

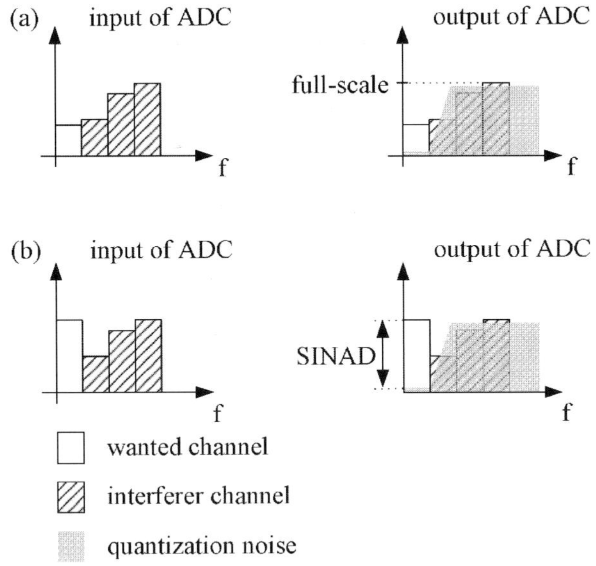

Figure 6.9: *Two input scenarios demonstrating the "waste" of SINAD in the conditioning channel of fig. 6.3.a*

Digital part: The digital filter needs to have a larger resolution and run at a higher sample rate $m_3 f_s$ than in the full-analog channel:

$$\frac{I_{\text{digital}}(\Sigma\Delta \text{ -channel})}{I_{\text{digital}}(\text{full-analog})} = \frac{m_3 \left[\log\left(DR\, q^2\right)\right]^3}{m_1 \left[\log SINAD\right]^3} \tag{6.9}$$

This discussion is illustrated for the same Bluetooth application as in chapter 5.

Example (continued from page 63:)
Applying eq. 6.6, with the implementation factor F equaling 1, by definition, the current increase due to the noise and distortion specification remains $\sim 10^5$. *The following topology is used:*

- *a conditioning ΣΔ ADC targeting SINAD=70dB is realized using* $m_3 = 32$, *in combination with a complex, feed forward loop filter of order* $L=5$ *(see section 9.2).*

Hence, the sample rate becomes 64MHz, which is only 2 times higher than for the conventional channel. Contrary to the multi-channel case, this sample rate is feasible both for the ΣΔ ADC and for the decimation filter. Filling out eq. 6.9, the power consumption in the decimation filter increases by a factor of 10 (assuming q equals one) as compared to the power in the digital part of the full-analog channel.

This example shows that a conditioning ΣΔ ADC can be a very power-efficient alternative to the conventional straightforward digitization of section 5.2.

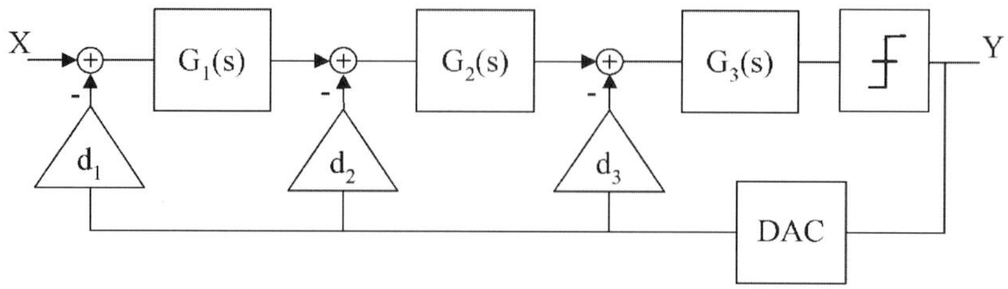

Figure 6.10: ΣΔ *ADC with a* 3rd*-order feedback loop filter*

Analog reference circuits

The very large *DR*-requirement on the ΣΔ ADC asks for a similar accuracy of the analog references (i.e. voltage or current reference and clock) used for the feedback DAC. Similarly, the bandwidth requirements of the reference circuits are related to the ADC sample rate. By consequence, these blocks remain challenging. In fact, this may be one of the most critical aspects of this particular conditioning channel. For instance, the realization in section 9.2 needs jitter below 0.01%.

In conclusion, *this architecture with signal-conditioning in the decimation filter is a highly-digitized alternative to the conventional channel. The power consumption of its ΣΔ ADC is comparable to that of the analog circuits in the full-analog channel. The decimation filter consumes more than the one in the full-analog channel but this increase seems acceptable; especially since this contribution shrinks with technology scaling anyhow. The accuracy required of peripheral reference circuits remains stringent.* An example implementation of this conditioning channel for a Bluetooth receiver is discussed in chapter 9.

6.3 Signal conditioning with a restricted filtering STF

It is shown that a ΣΔ ADC, based on a loop filter with nested feedback compensation, has a frequency selective STF. This property can be exploited to provide signal-conditioning in the ΣΔ ADC. This ADC is referred to as the "feedback ADC", as a shorthand notation. Fig. 6.10 shows an example for $N=3$. L_1 can be expressed as:

$$L_1(s) = -\left[d_3G_3(s) + d_2G_2(s)G_3(s) + d_1G_1(s)G_2(s)G_3(s)\right] \qquad (6.10)$$

and realizes the same noise transfer function as the feed forward ΣΔ ADC. Therefore, this topology achieves the same signal-to-quantization-noise-ratio. Its asset is in a better immunity to interferers because of the high-order transfer of L_0:

$$L_0(s) = G_1(s)G_2(s)G_3(s) \qquad (6.11)$$

In order to fully exploit this characteristic, variable gain control can be added in the input stage of the ΣΔ ADC. Then, the feedback ΣΔ ADC truly performs signal conditioning. First, the interferer immunity is investigated.

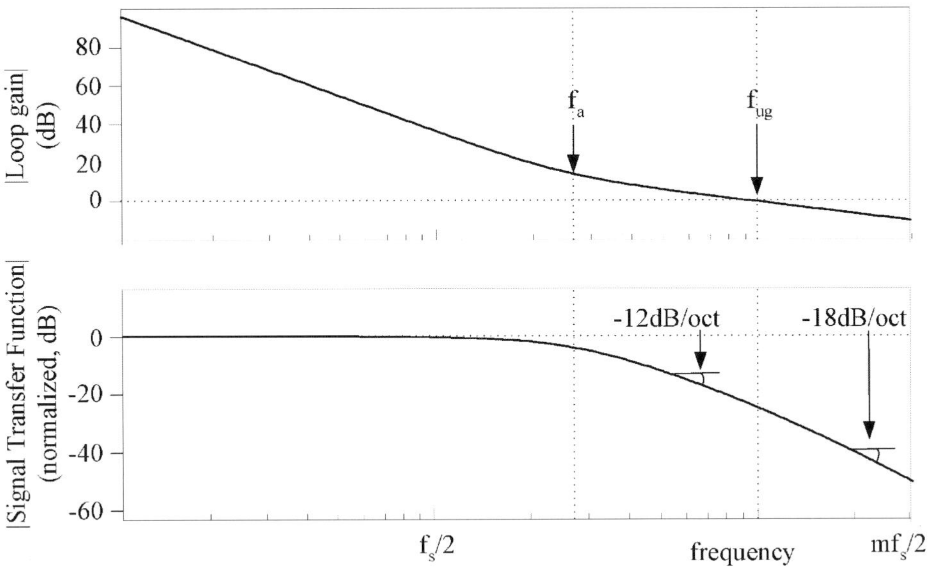

Figure 6.11: *Magnitude of the loop gain and the STF of the feedback ΣΔ ADC of fig. 6.10*

6.3.1 Interferer immunity

In the feed forward reference ADC the restricting limit on the allowable interferers is due to the available alias suppression and the stable input range. This is where the feedback topology excels.

Aliasing

In the feedback ΣΔ ADC, the interferer is subject to N^{th}-order filtering before being sampled in the quantizer; i.e. the order of L_0 in the nominator of eq. 6.3 equals N at the interferer frequency. This is different from the first-order filtering in the feed forward ΣΔ ADC. The noise-shaping at the alias frequency remains the same as in the feed forward case. Hence, the overall alias suppression function is of order $2N$ (instead of order $N+1$ for the feed forward implementation).

Stable input range

The loop gain $c\,L_1$ and the STF of the $3rd$-order feedback topology of fig. 6.10 are drawn in fig. 6.11. The frequency f_a indicates the transition from a high-order to a first-order slope in the loop gain; f_{ug} is the unity-gain frequency. Depending on the stability of the design f_a and f_{ug} approximate $mf_s/20$ and $mf_s/6$ respectively [25]. Using eq. 6.1,

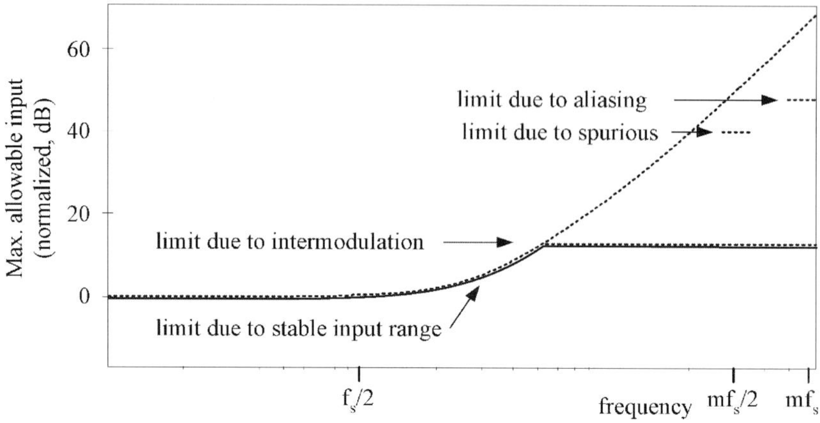

Figure 6.12: *Allowable input level for the feedback ΣΔ ADC of fig. 6.10*

eq. 6.11 and eq. 6.10 the STF of the feedback ΣΔ ADC equals:

$$\frac{Y(s)}{X(s)} = \frac{c\ G_1(s)G_2(s)G_3(s)}{1 + c\ [d_3G_3(s) + d_2G_2(s)G_3(s) + d_1G_1(s)G_2(s)G_3(s)]} \tag{6.12}$$

It features 2nd-order filtering for channels between f_a and f_{ug}, i.e. in the first-order region of the loop gain. Beyond f_{ug} the STF shows third-order filtering. Generalized, in case an Nth-order feedback loop filter is used, the STF features filtering of order $N-1$ for nearby interferers while far-off interferers experience filtering of order N. The stable input range is inversely proportional to the STF (eq. 6.5). *Notice, the $-3dB$-frequency of the STF corresponds to f_a. Hence, it can only be lowered at the expense of less noise-shaping. We will come back on this in section 6.4.*

Spurious responses

Spurious responses due to interferers around sub-harmonics of the sample frequency, are less likely because of the Nth-order filtering of L_0 before sampling.

Intermodulation distortion

In the feedback topology intermodulation distortion may be higher than what is predicted by eq. 4.12. Later on, it is analyzed that the signal swing on the various internal nodes of the feedback topology is much larger than in the feed forward topology. Hence, other distortion sources may become important as well.

Fig. 6.12 shows the various limitations (dotted lines) on the allowable input level of interferers over frequency for the example ADC of fig. 6.10. The solid line indicates the dominant limit. *For most frequencies, the restricting limit is due to implementation aspects (i.e. non-linear circuits) while the topology in itself is largely robust to interferers.*

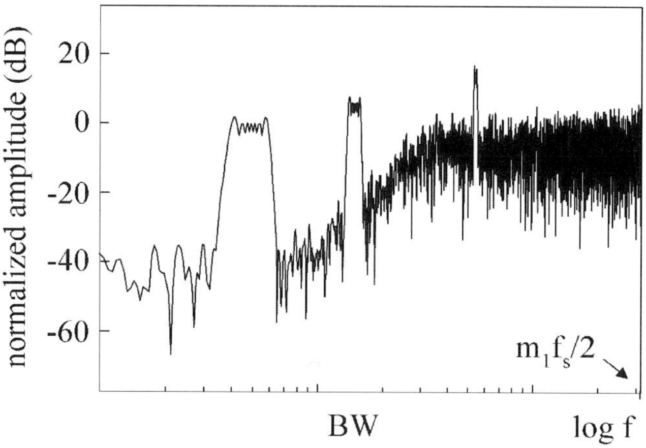

Figure 6.13: *Output spectrum of the feedback ΣΔ ADC*

Here, the distortion limit is indicated by a flat line, as an example. Depending on the position of the non-linearity in the loop, the curve may be very different (e.g. the example in section 9.3).

6.3.2 The conditioning channel

The increased interferer immunity and the filtering STF of this ΣΔ ADC *are optimally exploited in combination with variable gain control* of the input signal[8]. A gain setting between G_{MIN} and G_{MAX} is chosen such that the wanted signal is amplified to the maximum stable input level. Of course, interferers are amplified likewise. The allowable input level as characterized in fig. 6.12 should be tailored such that it can optimally accommodate the input spectrum of fig. 5.1, including the interferers.

Because of *the filtering STF and the integrated variable gain control the feedback ΣΔ ADC performs signal conditioning as depicted in fig. 6.3.b. The signal conditioning of the merged design is the same as that of the conventional cascade of filter, VGA and ADC in fig. 5.2.* Fig. 6.13 shows the output spectrum of the feedback ΣΔ ADC. Notice the relative attenuation of the interferers compared to the wanted signal. This feedback ΣΔ ADC with restricted conditioning is used in a similar way as the feed forward reference ADC in the sense that:

- the conversion bandwidth only accommodates the wanted signal;
- in the decimation filter, the remaining interferers are attenuated along with the quantization noise.

[8]The VGA functionality is easily integrated in the ΣΔ design for instance by making the input resistor switchable. Since resistors are reasonably linear over a wide bandwidth and voltage range, the VGA can precede the filtering. By consequence, the input stage (i.e. the first integrator) of the ΣΔ ADC does need to process the entire bandwidth of the input signal. Only, the *DR* of the signal is reduced because large wanted input signals are amplified less than in the full-analog channel. In conventional channels, the VGA normally follows the filter. Then, the *DR* and bandwidth requirements are stringent for the filter but become relaxed for the consequent stages.

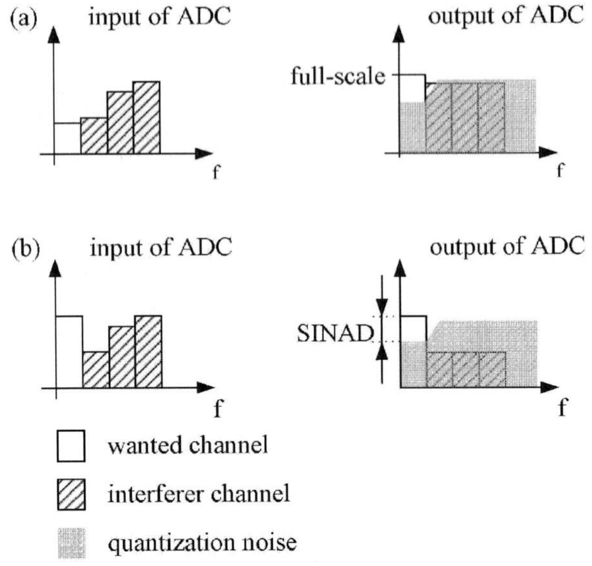

Figure 6.14: *Two input scenarios demonstrating the efficient use of the available SINAD*

The key difference compared to the reference ADC is in:

- the integrated filtering and variable gain functionality.

The latter characteristic implies the following important consequence:

- while the input-referred *DR* of the ΣΔ ADC remains large, at the output, the ΣΔ ADC only needs to provide a moderate *SINAD*.

It is shown later on that this results in a major overall power saving for the conditioning channel. First, this efficiency is intuitively demonstrated for the same example as in fig. 6.9

Example:
The same input scenarios as on page 82 are considered. In fig. 6.14.a the small wanted channel is amplified to the full-scale level of the ΣΔ ADC while the stronger interferer channels are attenuated to that level because of the filtering. In fig. 6.14.b the gain is reduced since the wanted signal is strong. The interferers are still attenuated. Because of the large wanted signal, the ΣΔ ADC noise can be relatively high. Hence, the quiescent current of the ΣΔ ADC can be adapted to the gain setting (i.e. lowered). By consequence, the SINAD of the ΣΔ ADC is used efficiently in both input scenarios.

Assuming the same amount of filtering and VGA as in the full-analog channel, this yields:

- an easy ΣΔ topology: i.e. the same low over-sampling factor, filter order and number of quantization levels can be used as for the ΣΔ ADC in the full-analog channel;
- an easy decimation filter: i.e. the same decimation filter as in the full-analog channel can be used with a low *ENOB*, sample rate and decimation factor;
- relaxed requirements on the DAC voltage, current or time reference: i.e. more noise and jitter on the DAC is allowed.

The penalty paid for this is in the control loop for the VGA. The complexity of this loop is comparable to that in the full-analog channel. Another penalty is in the higher implementation factor. This is discussed next.

6.3.3 Power/performance analysis

It is shown that a feedback topology inherently consumes more power than a feed forward implementation with the same specifications. On the other hand, the opportunity to integrate variable gain control lowers the average consumption of the ΣΔ ADC and reduces the specifications on the decimation filter and on the peripheral reference circuits. This yields an overall power saving.

The feedback ΣΔ ADC

In ΣΔ ADCs with a feed forward loop filter the current consumption of the second and higher integrators can be low since their noise and distortion is suppressed by the preceding gain. In the feedback topology, the contribution of the consecutive stages cannot be neglected because the unity gain frequency of the first integrator is much lower than in the feed forward case. This is due to the two reasons listed below.

- *Reverse order of integrator frequencies:* Though the feed forward and the feedback topology realize the same NTF and the same loop gain, the mapping of this function on the hardware is different. In the feed forward topology, the first integrator H_1 has the highest unity-gain frequency. The feed forward term $a_1 H_1(s)$ over-rules the other contributions at high frequencies and provides first-order behavior and thus (small-signal) stability. On the contrary, in the feedback ΣΔ ADC, this role is fulfilled by the inner loop $d_3 G_3(s)$. The same reasoning is valid for the other integrators: for stability reasons the feed forward topology can exploit an ascending order of unity-gain frequencies while the feedback topology needs a descending order.
- *Larger state-variables:* In a feed forward topology only the error signal is fed into the filter. The error signal primarily consists of the shaped quantization noise. In case a feedback loop filter is used the entire output spectrum -containing both the input signal and the quantization noise- is fed back to each internal node of the filter. By consequence, a strong compensating signal must be provided by each integrator. This is illustrated in the time domain in fig. 6.15. In order to maintain an acceptable signal swing at the integrator outputs; i.e. to keep the state-variables within a

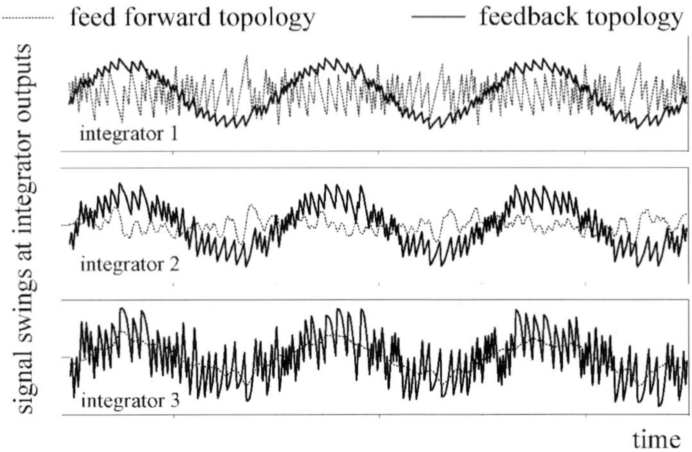

Figure 6.15: *Internal signal swings in a feed forward and a feedback topology*

common magnitude range, the unity-gain frequencies must be scaled down[9]. Additional information (for later use) on the frequency dependence of the internal signal swings is shown in fig. 6.16. It depicts the magnitude of the transfer function from the input of the ADC to the output of the integrators based on a linearized model of the ADCs in fig. 6.6 and fig. 6.10.

Especially for high-order loop-filters the unity-gain frequency of the first integrator(s) is significantly lower in the feedback topology than it is in the feed forward one. As the unity-gain frequency of the first integrator drops below the signal bandwidth, noise and distortion from the second and following integrators contribute increasingly to the overall *SINAD*.

Example:
Suppose, $P_n(G_2)$ represents a noise source at the input of block G_2 the equivalent input-referred contribution equals:

$$\left|P_{n,eq,G_2}(j\omega)\right| = \left|\frac{P_{n,G_2}(j\omega)}{G_1(j\omega)}\right| \tag{6.13}$$

Beyond the unity-gain frequency of G_1 the contribution of G_2 is amplified and can no longer be neglected. The same is true for its power consumption.

The above scaling effects and the consequent power increase in the various filter sections become very cumbersome in high-order filters. This is illustrated in section 9.3.

[9]Alternatively, a current-domain filter with a low-impedance load can be used. Since voltage and current are related via an impedance, a similar reasoning holds. In the current-domain solution, considerable power is often needed in order to realize a low noise impedance.

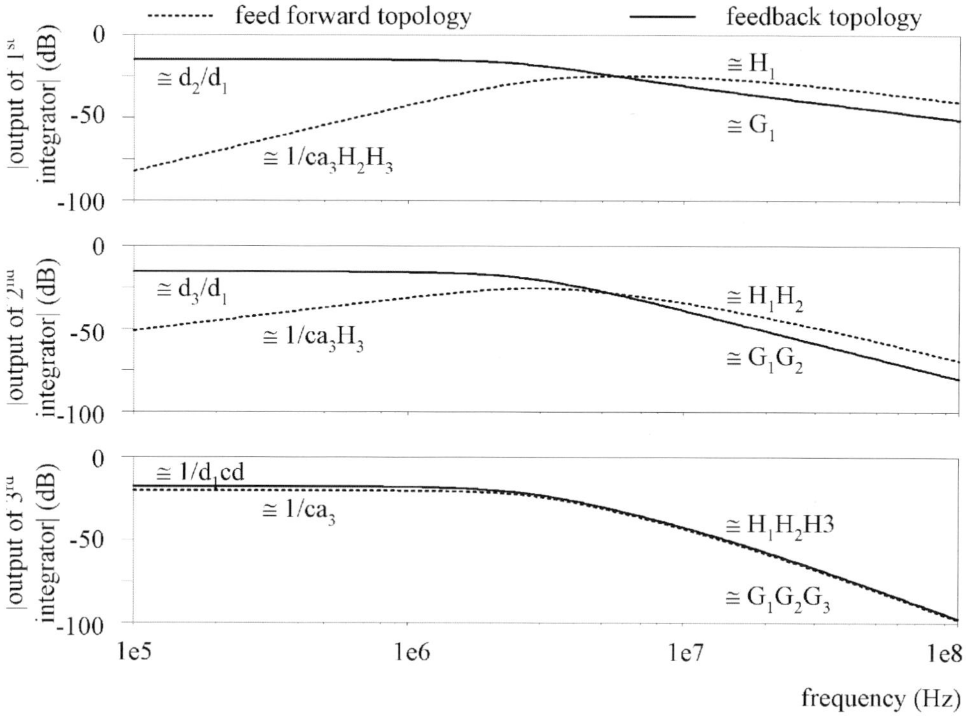

Figure 6.16: *Linearized transfer function from the input of the ADC to the internal nodes in a feed forward and a feedback topology*

The conditioning channel

Similar to fig. 5.5 the analysis focuses on the analog and the digital part of the channel separately.

Analog part: As discussed above, the feedback ΣΔ ADC consumes more than a feed forward implementation with the same specifications. Hence, eq. 6.6 applies with $F > 1$. For instance, it is argued on page 161 that $F \cong 1.5$ for that feedback design as compared to a conventional feed forward implementation.

On the other hand, the integrated, explicit signal conditioning (i.e. filtering of interferers and VGA) allows for dynamic adaptation of the quiescent current: if the wanted signal is strong, the input resistance of the ΣΔ ADC is lowered and the quiescent current can be decreased (see section 9.3). This adaptive biasing lowers the average current consumption. This benefit is hard to quantify since it depends on statistics of the received signal strength.

The integrated signal conditioning, with strong VGA, results in a moderate *SINAD* at the output of the ΣΔ ADC. Hence, the sample rate can be relaxed. Therefore, the required circuit bandwidth, is probably set by the interferer bandwidth instead of the sample rate. The requirement is then the same as for the input stage of the full-analog channel.

Digital part: Since the feedback ΣΔ ADC features filtering and VGA, essentially the same signal as in the full-analog channel is applied to the digital filter. Hence, the power consumption is as low.

Analog reference circuits

The accuracy required of the references is related to the *SINAD* at the output of the ΣΔ ADC, i.e. at the input of the DAC. Since this *SINAD* is moderate, the accuracy demand and the power consumption of the reference circuits are relaxed as well. *This is a major benefit* of maintaining some analog filtering and VGA, be it merged into the ΣΔ ADC. The same discussion holds for the bandwidth of these circuits.

In conclusion, this section shows the way to an elegant merge of the filter and VGA functionality in the ΣΔ ADC. The feedback topology has a worse implementation factor than the feed forward ΣΔ ADC of the previous channels. Still, because of its excellent immunity to interferers and the consequent opportunity to integrate VGA, the average current can be lowered. In addition, the digital part of the channel and the reference circuits are as easy as in the full-analog channel. It should also be kept in mind, that the filtering characteristic of the STF can only be optimized at the expense of the noise-shaping. An example implementation for a Bluetooth application is discussed in section 9.3.

6.3.4 Conditioning hybrid ΣΔ ADC

For completeness, it is mentioned that the hybrid topology of fig. 6.17 enables a trade-off between the characteristics of the feed forward and the feedback ΣΔ ADC. It is only briefly analyzed here. Starting from a feedback topology, feed forward coefficients c_i are added from the ADC input to (some) integrator inputs. Notice that these "feed forward" coefficients are different from those in the "feed forward ADC". In the latter case, the coefficients feed signals from the integrator outputs to the quantizer input. The filter functions L_0 and L_1 equal:

$$L_0(s) = c_1G_1(s)G_2(s)G_3(s) + c_2G_2(s)G_3(s) + c_3G_3(s) \tag{6.14}$$

$$L_1(s) = -\left[d_1G_1(s)G_2(s)G_3(s) + d_2G_2(s)G_3(s) + d_3G_3(s)\right] \tag{6.15}$$

Here, the coefficient c_i provide the compensating signal to lower the large low-frequency signal swing that is otherwise present at the integrator outputs (see fig. 6.15). As the signal swing drops the unity-gain frequency of the integrators can be scaled up again and

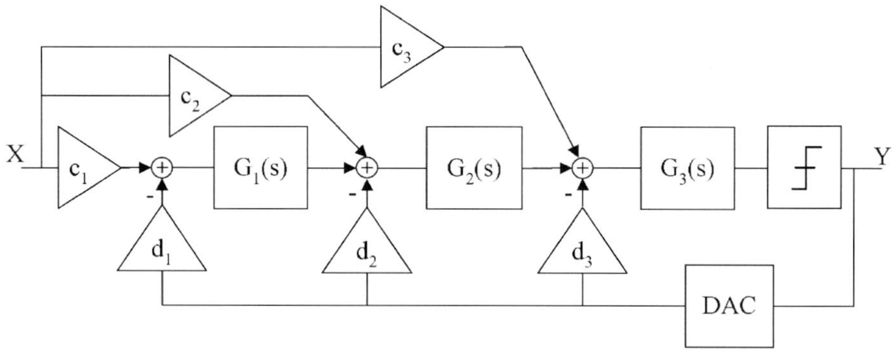

Figure 6.17: *Hybrid ΣΔ ADC*

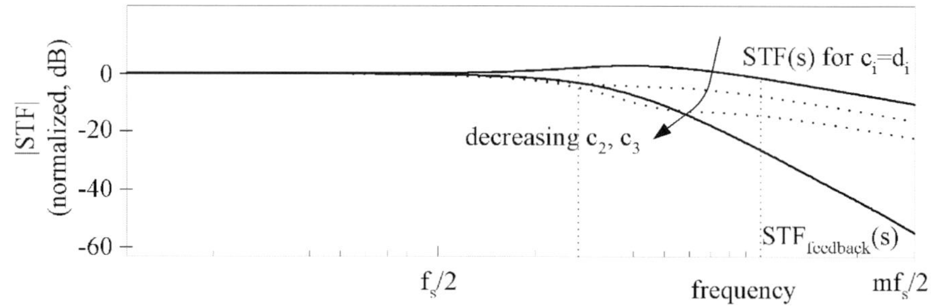

Figure 6.18: *Magnitude of the STF the hybrid ADC of fig. 6.17 with varying coefficient c_2 and c_3 simultaneously*

the input-referred noise and distortion from the filter sections G_2 and G_3 -as well as that of the coefficients c_2 and c_3- are reduced proportionally.

On the other hand, each feed forward path reduces the order of the STF by one. The STF of the ADC of fig 6.17 equals:

$$\frac{Y(s)}{X(s)} = c\,\frac{c_3G_3(s)+c_2G_2(s)G_3(s)+c_1G_1(s)G_2(s)G_3(s)}{1+c\,[d_3G_3(s)+d_2G_2(s)G_3(s)+d_1G_1(s)G_2(s)G_3(s)]} \tag{6.16}$$

In case the coefficients c_i equal the coefficients d_i (with $i = 1, 2, 3$) a flat STF is obtained within the unity-gain bandwidth of the loop. As the coefficients c_2 and c_3 are lowered, the compensation becomes incomplete and the STF evolves to that of the feedback topology. Fig. 6.18 shows the STF and the effect of varying the coefficients c_2 and c_3 jointly. The same loop gain as in the previous examples is realized. In the general case, the STF of a feedback ΣΔ ADC with a loop filter of order N, has N poles. In the hybrid topology, the added coefficients c_i introduce Z zeroes in the STF. As such, at high frequencies, the STF of the hybrid ADC has a slope of $N-Z$.

The alias-suppression is affected in a similar way: for each coefficient c_2, c_3, etc., that is added to the feedback topology, the order of the alias-suppression function is reduced by 1. If the coefficients c_i equal the coefficients d_i (with $i = 1, 2, 3$), the order of the alias-suppression function becomes $N+1$, just as for the feed forward ΣΔ ADC.

Hence, the coefficients c_i provide some flexibility in tuning the STF without changing the noise shaping. For a fixed NTF, tweaking these coefficients remains a compromise between the desired STF, the internal signal swings and the scaling of the unity-gain frequencies. *It can be concluded that, in terms of interferer immunity, the hybrid ΣΔ ADC enables a trade-off between the characteristics of the feed forward and the feedback ΣΔ ADC.* The further analysis is similar to that in the previous sections and the resulting power estimate reflects the above trade-off. Therefore, it is not discussed in more detail here.

6.4 Signal conditioning by unrestricted STF design

Instead of compromising between the power efficiency of the feed forward reference ADC and the robustness to interferers of the feedback ADC, both qualities are united in a ΣΔ ADC with unrestricted design of a filtering STF, from now on called the "filtering-feedback ΣΔ ADC" or FFB-ADC. *This terminology refers to the fact that at least one feedback path of the ADC comprises filtering.* Notice, this is not the case for the feedback ADC discussed before: there, the nested feedback paths consist of linear coefficients. Here, we present two specific implementations of a FFB-ADC. In the remainder, "FFB-ADC" refers to these examples, although, strictly speaking, it covers a wider range of implementations.

The presented implementations of the FFB-ADC are based on a conventional ΣΔ ADC with loop filter $H(s)$ to which a low-pass filter $H_{LPF}(s)$ and a compensating high-pass filter $H_{HPF}(s)$ are added. Here, a feed forward topology is considered for $H(s)$ in view of power-efficiency. Fig. 6.19.a and 6.19.b respectively show the parallel and the series configuration for the added filters (with subscript p to indicate the parallel or s to indicate the series configuration): Hence:

$$L_{0,p} = H_{LPF,p}(s) \cdot H(s) \tag{6.17}$$

$$L_{0,s} = H_{LPF,s}(s) \cdot H(s) \tag{6.18}$$

The added filters are complementary over the entire frequency range:

$$\begin{aligned} H_{LPF,p}(s) + H_{HPF,p}(s) &= 1 \\ H_{LPF,s}(s) \cdot H_{HPF,s}(s) &= 1 \end{aligned} \tag{6.19}$$

Therefore, the loop gain remains unaltered from that of the conventional ΣΔ ADC (i.e. the same L_1-characteristic is realized) and the noise shaping and the stability are the same as well.

In fact, the same idea can be applied to any of the discussed topologies. The feed forward ΣΔ ADC is chosen because of its superior power/performance relation. Also, instead of $H_{LPF}(s)$ any filter -or even any function- could have been used in principle. The only boundary condition is in the fact that the complementary function can be realized with a certain accuracy. The required matching between the complementary functions is discussed in section 9.4.

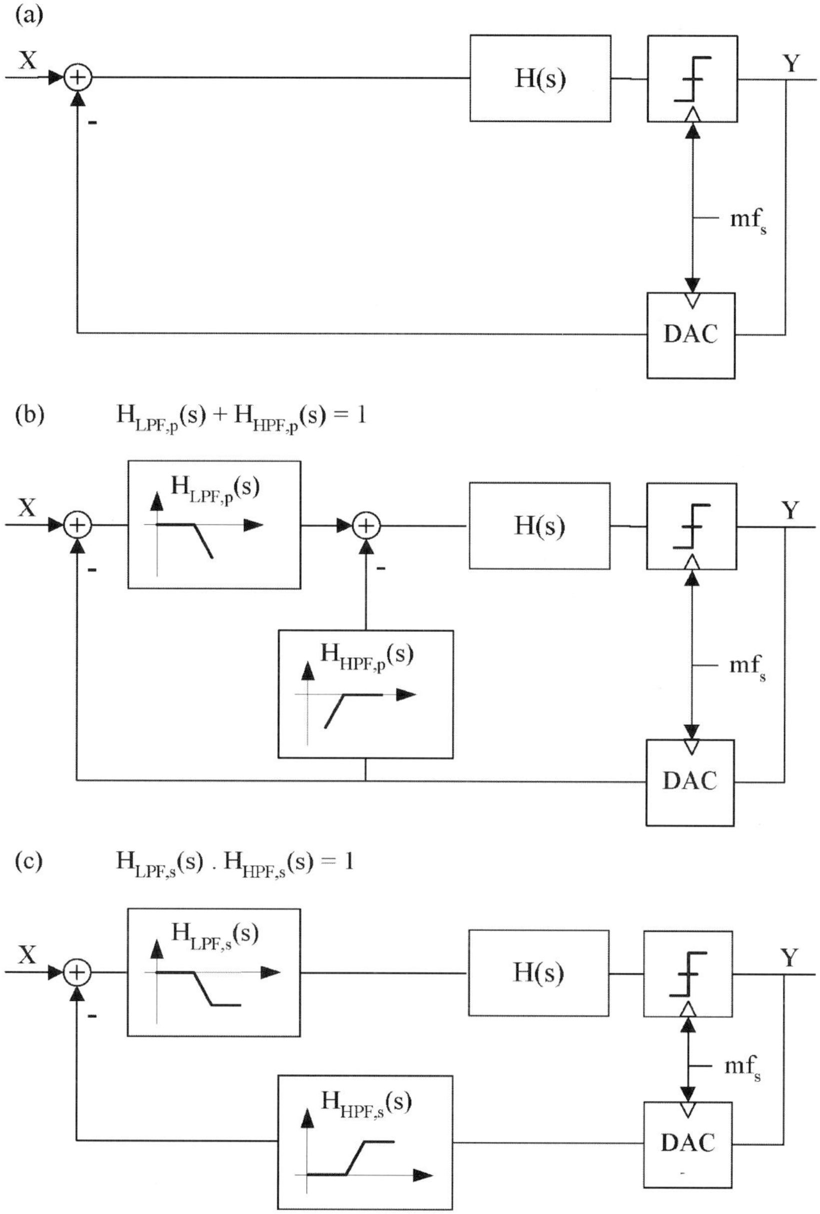

Figure 6.19: *Reference ΣΔ ADC (a) and ΣΔ ADC with explicit filtering added in a parallel (b) or a series (c) configuration*

The asset of the FFB-ADC, as compared to all previous topologies, is in the combination of an excellent power/performance ratio and a good immunity to interferers. Note that the functionality of the FFB-ADC is equivalent to that of a cascade of H_{LPF} and the reference ADC. Despite of requiring two filters (instead of one in the cascade), the FFB-ADC yields a more power and area efficient implementation. This is motivated in section 6.4.3 and illustrated with an example (page 174) in chapter 9.

6.4.1 Interferer immunity

The FFB-ADC has a better immunity to interferers than the feed forward ADC it is built on. This is analyzed next.

Aliasing

Suppose the order of $H_{LPF}(s)$ equals M and that of the conventional ΣΔ ADC equals N then the alias-suppression function is of order $M+N+1$. It can be calculated that both for the parallel and for the series configuration the alias suppression improves compared to the conventional feed forward ΣΔ ADC:

$$\frac{Y(\Delta f)}{X(mf_s-\Delta f)} \approx H_{LPF}(mf_s-\Delta f)\frac{H(mf_s-\Delta f)}{H(\Delta f)} \tag{6.20}$$

The improvement corresponds to the amount of low-pass filtering that is added.

Stable input range

The signal transfer function of the parallel and the series configuration are calculated first.

Parallel configuration:

$$STF_p(s) = H_{LPF,p}(s)\frac{c\,H(s)}{1+c\,[H_{LPF,p}(s)+H_{HPF,p}(s)]H(s)} \tag{6.21}$$

where c indicates the linearized quantizer gain as introduced on page 72. The DAC gain is assumed 1 for simplicity. Since the added filters are complementary over the entire frequency range:

$$STF_p(s) = H_{LPF,p}(s)\ STF_{\text{conventional }\Sigma\Delta\text{ ADC}}(s) \tag{6.22}$$

Series configuration:
In an analogy, it can be calculated that:

$$STF_s(s) = H_{LPF,s}(s)\ STF_{\text{conventional }\Sigma\Delta\text{ ADC}}(s) \tag{6.23}$$

The STF of the FFB-ADCs equals that of the feed forward ΣΔ ADC (fig. 6.7.b) multiplied by the low-pass characteristic of $H_{LPF,p}(s)$ or $H_{LPF,s}(s)$: the ΣΔ ADC explicitly

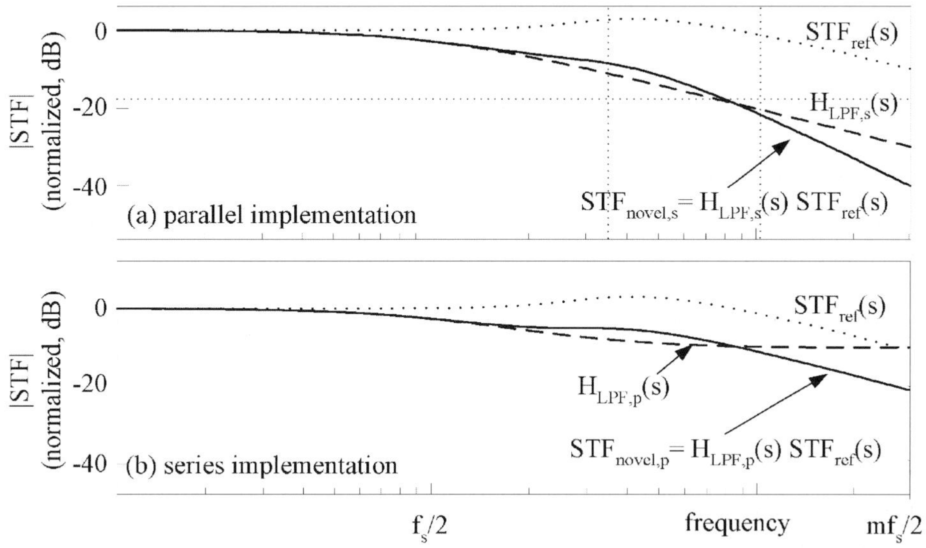

Figure 6.20: *Magnitude of the STF of the FFB-ADC of fig. 6.19.a and b (parallel and series configuration)*

provides filtering of the input signal towards the output. *Contrary to the previous ADCs the type, the order and the $-3dB$-frequency of the filtering can be chosen completely independent of the loop filter $H(s)$. This is a key asset of the FFB-ADC.*

The STF of the parallel and the series configuration are compared in fig. 6.20.a and b respectively. The same feed forward $\Sigma\Delta$ ADC as that of fig. 6.6 and the following first-order filter functions are assumed:

$$H_{LPF,p}(s) = \frac{a}{s+a} \quad \text{and} \quad H_{HPF,p}(s) = \frac{s}{s+a}$$
$$H_{LPF,s}(s) = \frac{s+b}{s+a} \quad \text{and} \quad H_{HPF,s}(s) = \frac{s+a}{s+b} \tag{6.24}$$
$$\text{with } a < b$$

The parameter a determines the $-3dB$-frequency of the low-pass filters. It is chosen to correspond to the signal bandwidth. The low-pass filter in the parallel configuration is designed to have a finite gain at high frequencies. This zero is provided by parameter b. It is chosen to correspond to 3 times the signal bandwidth.

The stable input range is proportional to the inverse of the STF (eq. 6.5) and thus increases as a function of frequency. It can be tailored by choosing the appropriate low-pass filter.

Spurious responses

Interferers are suppressed by the cascade of $H_{LPF}(s)$ and $H(s)$ before being sampled. Hence, they cause less correlated patterns than in the reference ΣΔ ADC.

Intermodulation distortion

Depending on the implementation of the FFB-ADC eq. 4.12 applies or other distortion sources contribute as well. This is further detailed in section 9.4.

As for the feedback ΣΔ ADC, it can be concluded that the topology of the FFB-ADC is largely immune to interferers. Circuit non-linearities are likely to limit the performance in most implementations.

6.4.2 The conditioning channel

The new conditioning channel can realize the same filtering as the feedback ΣΔ ADC of the previous section but it can even do better. Any filter $H_{LPF}(s)$ can be used as long as the complementary filter can be implemented with a reasonable accuracy. (It is shown in section 9.4 that the required matching is not very stringent.) Again, the filtering characteristic is optimally used in combination with variable gain control of the input signal. Then, signal conditioning corresponding to fig. 6.3.b is provided just as for the feedback ΣΔ ADC. In addition, the advantages listed on page 88 and the efficiency in terms of *DR* (illustrated in fig. 6.14) apply.

A key difference from the feedback ΣΔ ADC of section 6.3 is in an easier and more flexible implementation. This is discussed next.

6.4.3 Power/performance analysis

It is shown that the "filtering ΣΔ "-topology hardly consumes more than the feed forward ΣΔ ADC it is built on. Moreover, the combination with variable gain control of the input signal reduces the average consumption and relaxes the requirements on the decimation filter and the peripheral reference circuits.

The ΣΔ ADC

On an architectural level two blocks, i.e. the filters, have been added to the feed forward ΣΔ ADC. Practical implementations with only a small overhead in power consumption are possible though.

- *Contribution of H_{LPF}:* Part of the gain of the first integrator can be shifted into $H_{LPF}(s)$. (In the parallel configuration, the gain of $H_{HPF}(s)$ then needs to be increased by the same amount.) Alternatively, the order of $H_{LPF}(s)$ and the integrator sections can be interchanged. This will be demonstrated in section 9.4. Both measures result in sufficient gain in the input stage such that the noise and distortion of the following stages can be neglected.

- *Contribution of H_{HPF}:* A simple, i.e. first-order, high-pass filter can easily be integrated in the DAC with passive components only. An alternative solution is given in section 9.4.

There, an implementation of the FFB-ADC is presented. It achieves $F \approx 1.1$ (see page 174).

The conditioning channel

Again, the analog and digital part of the channel are pair-wise compared to the corresponding parts of the full-analog channel.

Analog part: The implementation factor (with respect to noise and distortion) almost equals 1 and, thus, is much more favorable than for the feedback ΣΔ ADC or the hybrid ΣΔ ADC. Because of the integrated signal conditioning, dynamic biasing is possible. It can be programmed along with the gain setting. Consequently, the average current can be lower than in the full-analog channel.

Bandwidth-wise, the requirements are comparable to those for the input stage of the full-analog channel. In this respect, the discussion is the same as that for the feedback ADC of section 6.3.

Digital part: The discussion on the digital part of the feedback ΣΔ ADC on page 92 is valid here too.

Analog reference circuits

The requirements on the references are as easy as for the full-analog channel and for the feedback ΣΔ ADC (see discussion on page 92).

In conclusion, *the FFB-ADC combines the advantages of the highly-digitized channels based on the feed forward ADC with signal-conditioning in the decimation filter (section 6.2) and on the feedback ΣΔ ADC with a restricted filtering STF (section 6.3). It has a comparable, low power consumption as a full-analog channel of section 5.1. The filtering STF and the implementation of VGA allow a lower average consumption in the ADC, an easier decimation filter and relaxed requirements on all references just as for the feedback ADC. Contrary to the feedback ADC, the filtering STF can be optimized completely independent of the NTF.* An example implementation, again targeting Bluetooth specifications, is presented in section 9.4.

6.5 Comparison of conditioning ADCs

The various conditioning ADCs are compared in terms of filter topology. This leads to a better understanding of the constraints and the opportunities of each architecture. Next, the flexibility and power consumption of the various ADCs is compared in order to distill guidelines for a power-efficient and flexible design.

6.5.1 Comparison of topologies

Although the presented ΣΔ ADCs all obey the universal model of fig. 6.4, they do differ in the mapping of the functions L_0 and L_1 on a filter topology. In addition, they may also realize a different L_0-function. In view of optimal noise shaping, the function L_1 is the same for all ADCs, though.

In the comparison, the path from the input of the ADC to the quantizer -described by L_0- is referred to as the "forward path". The path from the output of the ADC to the input of the quantizer -described by L_1- is called the "return path".

Signal conditioning in the decimation filter (section 6.2)

Because of the interferer immunity of continuous-time ΣΔ ADCs, this technique is possible in combination with any of the discussed ΣΔ ADCs. In case of a feed forward implementation, this is the single opportunity for signal-conditioning in the "broader" ADC. In fig. 6.21.a, the universal model of fig. 6.4 is translated to this implementation topology. All filter sections are in the forward path and only one return path is implemented. Since $L_0 = L_1$, the STF is more or less flat over a wide frequency range (eq. 6.1). It cannot be used for filtering.

Signal conditioning with restricted filtering STF (section 6.3)

These conditioning ADCs are captured by the topology of fig. 6.21.b. Still, all filter sections are in the forward path. However, multiple forward and return paths are available. As such, $L_0 \neq L_1$ becomes possible, enabling $|STF| < 1$ for interferers channels. The nested feedback paths are essential for obtaining the filtering STF. The forward paths are optional; e.g. in the hybrid implementation of section 6.3.4 they are introduced for reducing the internal signal swings. Despite of the additional forward and return paths, in the feedback and the hybrid implementations, all filter sections of the return path are shared with the forward path. By consequence, the realizable STF is constrained by the desired NTF.

Signal conditioning by unrestricted STF design (section 6.4)

This is the single topology where the return path(s) include(s) filter sections that are not present in the forward path: in fig. 6.21.c, the filters F_1 and F_2 are added in the return path. F_1 and F_2 introduce degrees-of-freedom to design for a filtering STF. Notice, this is a generalized topology. In the ADCs presented in section 6.4, either F_1 is a short or F_2 is an open connection. The additional forward path H_2 is absent, there.

In conclusion, the nested feedback enables a filtering STF. However, the fact that all filter sections are shared between the forward and the return path, restricts the achievable

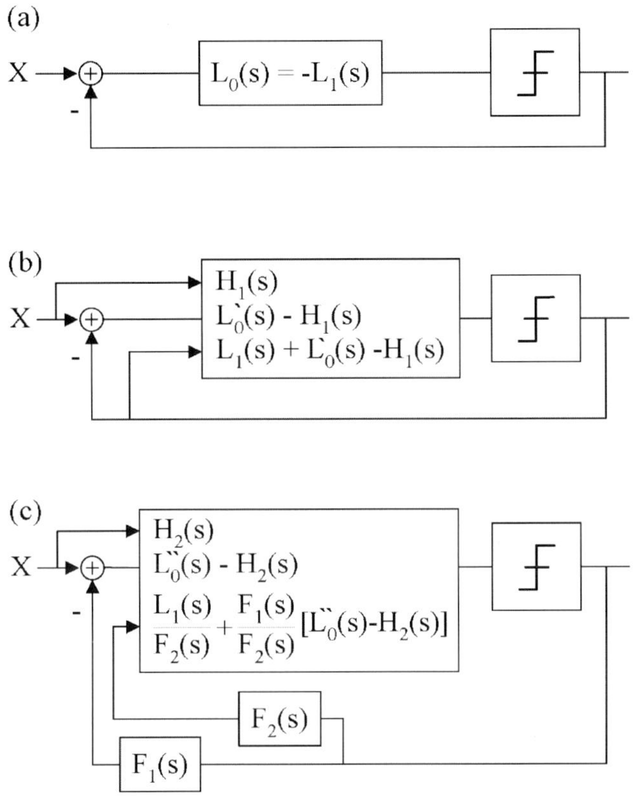

Figure 6.21: *ADC topology enabling: signal-conditioning in the decimation filter only (a), signal-conditioning with a restricted filtering STF (b) and signal-conditioning by unrestricted STF design (c)*

STF. On the contrary, adding filter sections in the feedback allows for unrestricted filtering STF design.

6.5.2 Flexibility

The comparison in this and the following section is summarized in table 6.2. The rating of the various channels is motivated below.

Flexibility is not considered as a synonym for digitization here. Of course, digitization helps but analog circuits feature several degrees of flexibility too. For example, the conventional architecture, with a cascade of analog sections, needs many control loops for tuning of gain, offset or filter frequencies. In order to accommodate the remaining tolerances, the circuits need to be designed with sufficient margin. As a consequence, it is rated rather "inflexible".

Table 6.2: *Comparison of the various channel topologies*

Channel	P-efficiency	Flexibility
full-analog channel (sec. 5.1)	0	−−
full-digital channel (sec. 5.2)	−−	++
con.[a] ΣΔ ADC with:		
feed forward reference ADC		
(sec. 6.2: con. in the decimation filter)	+	+
feedback ADC		
(sec. 6.3: restricted, filtering STF)	-	+
FFB-ADC		
(sec. 6.4: unrestricted filtering STF)	++	+

[a]con.=conditioning

The conditioning ΣΔ channels featuring filtering and VGA (i.e. the feedback ADC, the hybrid ADC and the FFB-ADC) need less accuracy for the analog blocks. Though any of the ΣΔ ADCs may need tuning of the time-constants in case of large process spread, the tuning is not critical. It must only guarantee that the noise shaping doesn't start within the conversion bandwidth. Furthermore, these channels benefit from the overall feedback of the ΣΔ ADC: the loop operation reduces the accuracy and performance demands of the individual filter sections. Hence, the ΣΔ channels are all reasonably flexible in terms of scaling, robustness, etc.

For some applications, the feed forward ΣΔ ADC, with signal-conditioning in the decimation filter, may be interesting because it provides the full signal *DR* to the digital domain. Obviously, the full-digital architecture offers the highest flexibility. It is the single architecture that enables full-digital selectivity, both in terms of amplitude and in terms of frequency.

6.5.3 Power consumption

The conventional analog channel is taken as a reference. As previously discussed, the conditioning feed forward ΣΔ ADC is likely to consume less: the quiescent current of its input stage is comparable to that for the first stage of the full-analog channel while the quiescent current of the following stages can be lower because of the overall feedback.

Likewise, for the feedback, the hybrid and for the FFB-ADC, the quiescent current of the input stage is comparable as well. However, the overhead of the remaining stages is different. For the feedback implementation it is largest. For the FFB-ADC, the overhead is negligible in case the added filters are of low-order. All architectures with a filtering

STF allow the integration of VGA. Consequently, their quiescent current can be adapted to the gain setting: if the input signal is large, the bias current can be lowered. This reduces the average power consumption.

Finally, the flexibility of the full-digital architecture comes at the penalty of a significant power increase. Foremost, the ADC is very difficult and it may run into various sample rate related limitations (section 3.4). In addition, the accuracy demands on the analog references become stringent and even the digital filter becomes challenging because of the high clock frequency.

6.5.4 Guidelines

From the discussion in this chapter and the comparison above, guidelines for the design of a low-power and flexible conditioning channel are distilled:

- in view of distortion and noise, all channel topologies require about the same quiescent current for the input stage. However, the "overhead" current of the consequent stages can be lowered by applying overall feedback and providing high gain in the first stage. As such, a conditioning channel based on a ΣΔ ADC -preferably with a feed forward loop filter- is advised;
- the signal conditioning should happen early in the channel in order to reduce the bandwidth and resolution requirements in the following of the channel and in the reference circuits. In the conditioning ΣΔ ADCs, this is achieved by the dedicated design of L_0;
- the sample rate of the ΣΔ ADC must be kept as low as possible in view of power consumption in the ΣΔ ADC, power consumption in the digital part and accuracy and bandwidth requirements on the reference circuits;
- digital control -e.g. of the VGA- can be implemented to reduce the average power consumption of the analog blocks;
- a wise balance between flexibility and power efficiency of the conditioning channel should be pursued. For example, in the conditioning ΣΔ ADCs presented in this chapter, the selectivity still happens -to a large extent- in the analog domain but the burden of tuning and calibration of analog inaccuracies is lifted. This results in a reasonably flexible, but especially very power-efficient channel.

6.6 Conclusions

In this chapter, the terminology of "conditioning ΣΔ ADCs" and of a "filtering-feedback ΣΔ ADC" -constituting a specific implementation form of a conditioning ΣΔ ADC- has been introduced. The ΣΔ ADC is at the heart of an alternative signal conditioning channel.

An alternative digitization strategy is presented: instead of replacing analog functions by digital processing, these functions can be integrated into a "conditioning ΣΔ ADC".

This strategy is slightly less flexible but much more power-efficient than the straightforward digitization of section 5.2. Compared to the full-analog channel of section 5.1, it is likely to result in a lower power consumption for the channel. In addition, it relaxes the accuracy requirements on the analog reference circuits.

The key pillar of the "conditioning sigma-delta ADC" is a rather under-exploited characteristic of continuous-time sigma delta A/D converters, namely their inherent immunity to interferers.

This immunity enables signal-conditioning in the decimation filter, e.g. using the feed forward reference ADC, or even in the ADC, e.g. using an ADC with nested feedback in the loop filter. As an improvement on the latter, restricted and power in-efficient form of signal conditioning, the FFB-ADC is presented. It features excellent power efficiency in combination with explicit and flexible conditioning.

This is illustrated with example implementations in chapter 9.

Chapter 7

Digitization of the inter-die interface

In the previous chapter, alternative solutions for the digitization of the conditioning channel have been presented. In practice, the conditioning channel may extend over multiple dies, possibly in a different technology. The ADC can be integrated with the analog part or it is put on the digital die. Depending on this partitioning choice, the inter-die interface will be digital or analog. It is shown in this chapter that the digitization of the conditioning channel leads to a digitization of the inter-die interface as well. This is depicted in fig. 7.1.

The chapter starts with some general considerations on the choice of the interface. Next, possible IC partitioning scenarios for a ΣΔ based conditioning channel are compared with respect to power consumption in the associated interface. The analysis on power consumption supports the on-going digitization of the inter-die interface for high-resolution conditioning channels.

7.1 Considerations

In case of a multi-die solution, the choice for an analog or a digital interface can be determined by a number of factors, a.o.:

- *technology choice for ADC/cost:* The ADC can either be integrated in an advanced CMOS technology or a technology with high-performance analog capabilities. Often, the choice for one or the other is cost-driven.
- *electro-magnetic interference:* In case of a digital interface the communication can be a source of electro-magnetic interference affecting the performance of other circuits. The opposite is true for an analog interface: it is susceptible to interference from other sources. A differential implementation is therefore preferable in all cases. For a digital interface low-swing operation is favored while for an analog link a large signal is preferred.

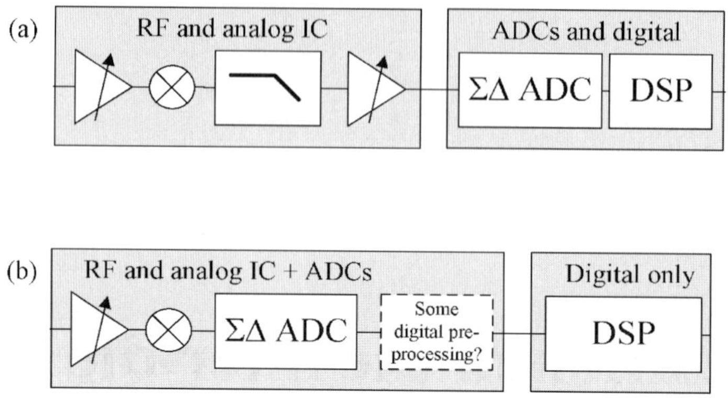

Figure 7.1: *Analog interface for a full-analog conditioning channel (a) and digital interface for a full-digital channel*

- *pin count:* An analog, differential interface requires two pins per signal path. The same is true for a digital, differential interface. Likely, though, it needs many more signal paths because of the number of parallel output bits. These bits can be serialized to reduce the pin count. Next to the serializing/de-serializing this also requires one additional signal path to transmit a synchronization signal. In addition, a digital interface requires transmission of the clock, unless clock recovery is implemented.

- *standardization:* In an industrial realization, compatibility of the analog die with various digital ICs and vice versa is of major importance. As a consequence, the inter-die interface is being standardized for various applications. This argument is illustrated for the example of a GSM chip set.
  Partitioning of a GSM chip set: Fig. 7.2.a shows a typical chip partitioning for the GSM chip-sets of the 1990's: it has an analog interface between the RF IC and the digital baseband with the ADCs and DACs. Meanwhile, a consortium of companies is defining a digital interface standard "DigRF" for the next generation of GSM solutions (see fig. 7.2.b and [86]). Similarly, standards like a.o.'MiPi" [87], "JC61" [88] and "BlueRF" are being developed.

- *power/performance of the interface:* A last argument for choosing a digital or an analog interface is in their different power/performance relation.

This is analyzed next. The analysis of the interface is conducted from the angle of digitization of the channel. Hence, it is of a limited nature.

7.2 Power in the interface

In a ΣΔ based conditioning channel the IC partitioning can be such that the inter-die interface is either analog (before the ADC), digital after decimation or digital but before decimation (see fig. 7.3). The power associated with these three possible interfaces is

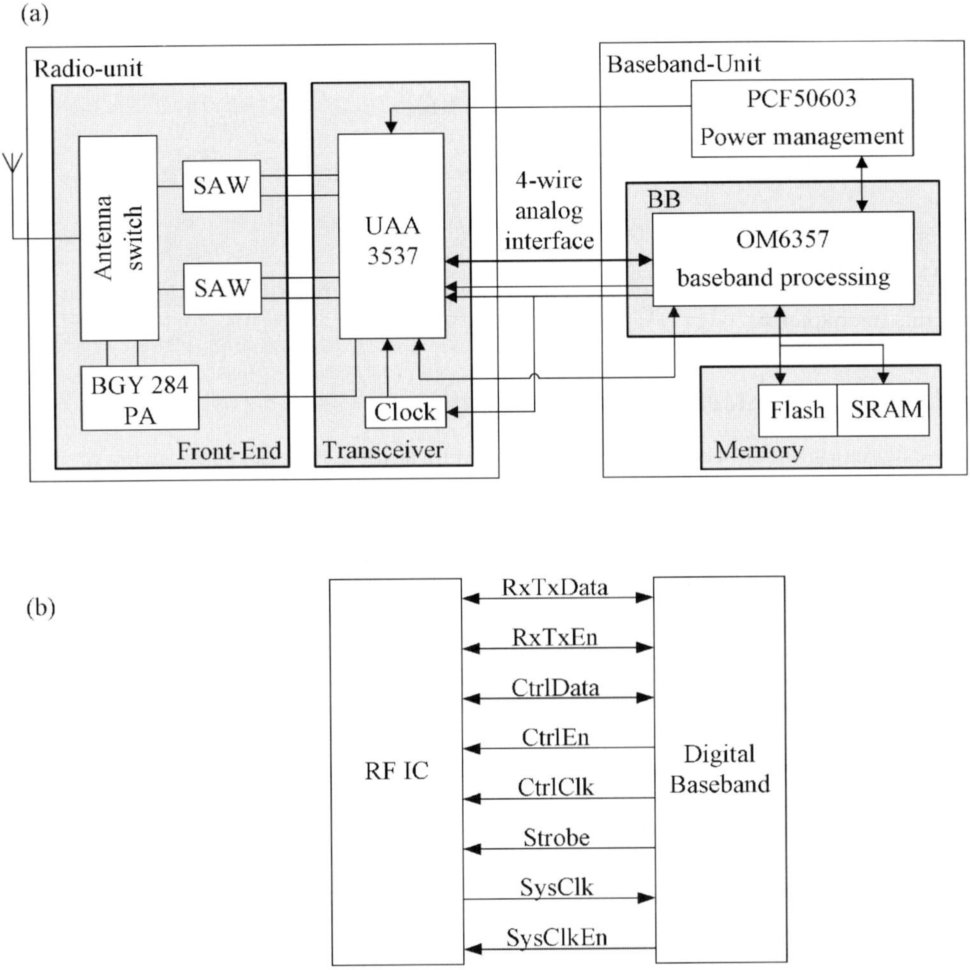

Figure 7.2: *Conventional GSM solution with analog inter-die interface (a) and proposed digital interface standard "DigRF" (b) for next generation, highly-digitized GSM solutions [86]*

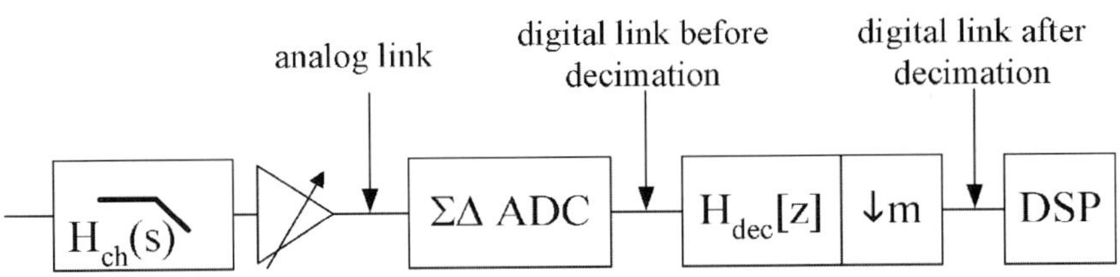

Figure 7.3: *Possible IC partitioning for ΣΔ based conditioning channels*

discussed next. It is based on the power calculations of analog and digital data interfaces in appendix E and in [89].

7.2.1 Analog interface

From the calculations in appendix E the power consumption of an analog interface follows:

$$P \sim 2^{2ENOB} BW \tag{7.1}$$

The result is similar to the power/performance relation defined for analog conditioning circuits in chapter 4.

7.2.2 Digital interface after decimation

The calculations in appendix E distinguish between a full-swing and a low-swing or slew-rate controlled interface.

The dissipation in a full-swing interface is dynamic only, i.e. it is solely due to the charging and discharging of a load capacitance. Simplifying eq. E.7 the power consumption in a full-swing interface follows:

$$P \sim ENOB\, f_{sw} \tag{7.2}$$

where frequency f_{sw} is defined as the inverse of the average number of *0 – 1* transitions in the data. It is considered proportional to the sample frequency. For simplicity, it is assumed that all bits are "effective" and, thus, *ENOB* is filled out. For decimated data, the difference between the number of bits and the effective number of bits is small.

Alternatively, a low-swing interface can be implemented for interference reasons. In that case, part of the dissipation is dynamic and follows eq. 7.2. Another part is static: it is due to the DC biasing current and is independent of *ENOB* and f_{sw}.

Hence, it can be concluded that *the current consumption of a digital interface is proportional (in case of a full-swing interface) or less than proportional (in case of a low-swing interface) to the data rate,* i.e. the product of *ENOB* and f_{sw}.

7.2.3 Digital interface before decimation

A ΣΔ ADC exchanges resolution in amplitude (i.e. *ENOB*) for resolution in time (i.e. f_{sw}). This exchange is not linear. By consequence, a ΣΔ ADC has a higher output data rate than what is essentially required in view of the *ENOB* and the signal bandwidth; *i.e. not all output bits are "effective"*. This is true for ΣΔ encoding in general, but especially for single-bit ΣΔ encoding:

- single-bit quantization adds a considerable amount of quantization noise (much more than high-resolution Nyquist A/D conversion)

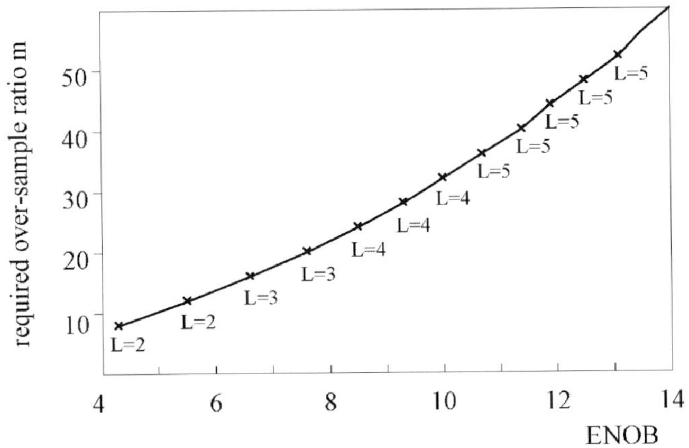

Figure 7.4: *Required over-sampling ratio m as a function of the ENOB target and the effect of increasing the filter order L*

- single-bit ΣΔ encoding reduces the maximum modulation depth of the digital output by a factor of two compared to the digital full-scale power.

In case the IC partitioning is such that the ADC and the decimation filter are on separate dies, *power is "wasted" for the communication of quantization noise in between the dies.*

In order to express the data rate at the output of the ΣΔ ADC in terms of *ENOB* and signal bandwidth, eq. 3.1 is used. The following relation between the over-sample ratio m and the effective number of bits *ENOB* can be derived:

$$m \cong 2^{\frac{ENOB}{L}} \frac{\pi}{(2L+1)^{\frac{1}{2L+1}}} \tag{7.3}$$

with L indicating the order of the loop filter. This suggests that the required over-sampling increases exponentially as the *ENOB*-target grows. In practice, though, m is a much weaker function of *ENOB*, because:

- L is a function of m: as m increases, a higher-order loop filter is normally implemented as well
- the function $(2L+1)^{\frac{1}{2L+1}}$ only varies between 1 and $e^{1/e}$ for positive values of L. Hence, this variation can be neglected.

The above is clarified with the curves in fig. 7.4. The plotted relation is based on C-simulations of a bank of available filters for various values of m. The combination yielding the highest *ENOB* is plotted. Alternatively, eq. 3.1 could have been used but, as mentioned on page 23, that estimate may be too optimistic because it doesn't take into account stability related limitations.

The switching frequency of the digital link is proportional to m. Filling out relation eq. 7.3 in eq. 7.2 the dynamic dissipation of a digital inter-die interface with ΣΔ modulated data becomes:

$$P \sim 2^{\frac{ENOB}{L}} BW \tag{7.4}$$

Finally, note that an important asset of a digital link with bitstream data is in its robustness for bit-errors in the code. Contrary to a multi-bit code, in the bitstream code all bits are of equal weight and any bit-flip causes only a small error. This remark links to the historical overview as described in section 3.1.

7.2.4 Comparison

For the three interfaces discussed above, the current consumption versus *ENOB* is plotted in fig. 7.5. Since the above relations only predict a proportionality, the absolute position of the curves depends on the implementation, the various specifications (for example the load capacitance, the signal swing, etc.) and the margins taken in view of interference. Here, eq. 7.1 and 7.2 are normalized such that these graphs intersect at $ENOB = 10$ as an example. It is observed that, *for low-resolution -probably large bandwidth- applications an analog interface consumes least while in case of a high-resolution -probably low bandwidth- application a digital interface is favorable.*

The curve of eq. 7.4 (corresponding to the $\Sigma\Delta$ encoded data) is fixed relative to that of 7.2 (corresponding to the decimated date). As such, the intersection of eq. 7.1 with eq. 7.4 lies at a higher *ENOB*-value: assuming both types of digital interfaces are implemented on the same hardware, *the interface before decimation consumes more than the one after decimation because of the $\Sigma\Delta$ encoded data.* The difference is less dramatic though than what is suggested on first sight when comparing eq. 7.2 and eq. 7.4 (see discussion on m as a function of *ENOB*). *For high-resolution channels, the interface before decimation remains attractive compared to the analog interface.*

Depending on the application, *the decimation filter may also include scaling of the word-length* (i.e. digital VGA: the input-referred *DR* can be reduced to the *SINAD* required for further processing). *In that case, a digital interface with decimated data becomes significantly more attractive* because of the further reduction in data rate.

7.3 Application to the conditioning channels

The above findings are applied to the channels discussed in chapter 5 and 6. A decisive conclusion on the choice for the interface should include the application (for example high- or low-resolution) and the considerations listed on page 105. Here, a high-resolution, moderate bandwidth application, as for example wireless communications, is assumed and power consumption is the dominant decision criterion.

Under these assumptions, the following guidelines are distilled.

- *Choose an analog interface for the conventional channel of section 5.1:* since a lot of analog conditioning is applied only a limited *DR* is required and an analog interface consumes least.

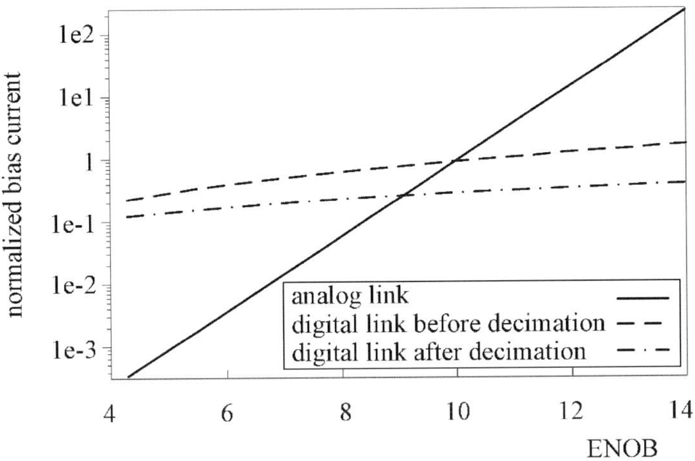

Figure 7.5: *Current consumption in the inter-die interface as a function of the number-of-bits (with arbitrary intersection for ENOB = 10)*

- *Choose a digital interface with decimated data for the multi-channel architecture of section 5.2:* this is the best choice for a high-resolution link. Moreover, the decimation filter may include channel selection and reduce the signal *DR* to the *SINAD* needed for digital processing.

- *Choose any digital interface for the conditioning ΣΔ architectures:* since the ADC in these architectures includes some signal conditioning its output data rate is moderate and a digital link remains favorable. Because of the robustness of ΣΔ -encoded data to bit-flips, the digital interface before decimation may be preferred..

In view of power consumption, the general advice is to perform the signal conditioning before interfacing to a following die. By consequence, *as the signal conditioning is being digitized, the interface becomes digital as well.*

7.4 Conclusions

For low-resolution, large bandwidth applications, an analog inter-die interface is preferred for power reasons. For high-resolution, low bandwidth applications, a digital interface consumes less.

In an analogy to the results for signal conditioning (chapter 4), it is found that the power consumption of an analog data interface is exponential in *ENOB*. Power consumption of a digital interface is linear or less than linear in *ENOB*, at least, in case decimated data is assumed. In case the data is ΣΔ encoded, the power consumption of the digital interface shows a weak-exponential dependence on *ENOB*.

For highly digitized conditioning channels, a digital inter-die interface should be used in view of power consumption.

Chapter 8

Highly analog and highly digital channels for FM/AM radio

This chapter illustrates the theory in chapter 5 on highly analog and highly digital conditioning channels. A dual-mode conditioning channel for use in an FM/AM car radio receiver is presented. Elaborating on the rest of the book, this channel includes some IF blocks in addition to the baseband functionality. In FM mode, the signal conditioning is highly analog (as described in section 5.1). In AM mode, the same conditioning channel is re-used. As such, a sufficiently large bandwidth becomes available for multi-channel A/D conversion (as described in section 5.2). However, the multi-channel aspect leads to very challenging linearity requirements. This becomes obvious from the design of a CMOS VGA around the IF frequency and from the design of the ΣΔ ADC performing the multi-channel A/D conversion.[1]

8.1 System

Although FM and AM radios have been around for decades, there is a clear need for further integration and digitization of the system. This is especially true for a car radio system. Both cost- and performance-wise this is a very demanding application. Digitization of signal processing and conditioning brings the following advantages a.o. [90]:

- *higher quality:* DSP algorithms can be used to reduce the effect of a.o. ignition noise and interferers. Software-controlled adaptive bandwidth of digital channel-select filters improves the signal quality and fidelity: this bandwidth can be chosen depending on reception conditions such as the strength of neighboring channels;

[1] This chapter heavily relies on work by Eric van der Zwan. He derived the specifications on the ADC and defined the ADC architecture. He also took the lead in the publication of [2], of which some parts have been re-used here. The implementation of the ADC -though heavily inspired by previous work of Eric a.o. [7]- was performed by the current author. In addition, the analysis, the evaluation in the present context and the benchmark are original work, just as all VGA-work.

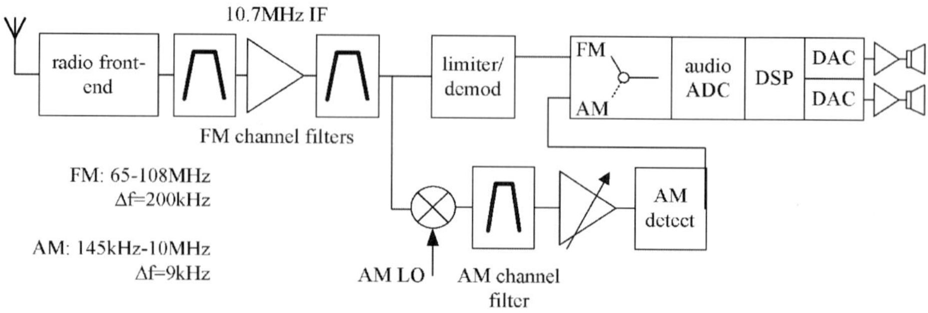

Figure 8.1: *FM/AM radio with analog demodulation*

- *more features:* Many aspects of the radio functionality and performance can be parameterized and can be made user-controllable;
- *more flexibility:* Quality and features can be tailored to any market or customer because of programmable, digital filtering and processing.

Below, a radio with analog signal conditioning and demodulation is described first. Next, a radio with digital demodulation is presented. In FM-mode, the signal conditioning remains highly analog but in AM-mode, multi-channel A/D conversion and full-digital selectivity are implemented.

8.1.1 Conventional radio with analog demodulation

Fig. 8.1 shows a block diagram of a FM/AM radio receiver with analog demodulation. Depending on the continent, FM radio operates in between 65-108MHz. AM channels, including LW, MW and SW frequency bands range from 145kHz to 10MHz. The radio front-end converts the FM or AM antenna signal to an intermediate frequency (IF) of 10.7MHz. Analog FM demodulation takes place after 200kHz channel selection at the IF. The channel-selection may consist of more than one external filter; e.g. the high-end car-radio application described in [91] needs *four external filters*. Typically, these are high-quality, surface acoustic wave filters (SAW) possibly combined with a coil. AM signals are processed by a double conversion receiver using the same 10.7MHz frequency as first IF. After mixing to a second IF (for example 450kHz), 9kHz channel filtering (by another expensive, external filter) and amplification by a VGA, analog AM demodulation takes place. The demodulated FM signal or the AM audio signal is digitized in the audio signal processor IC. This signal processor may provide functions like interference absorption, stereo decoding, radio data system (RDS) decoding and audio controls such as volume, balance, tone control and dynamics compression. At the output, the digital signals are converted to analog signals, which connect to the power amplifiers.

Notice that this radio receiver has full-analog selectivity: even the demodulation of AM and FM signals happens in the analog domain. On the other hand, a lot of digital

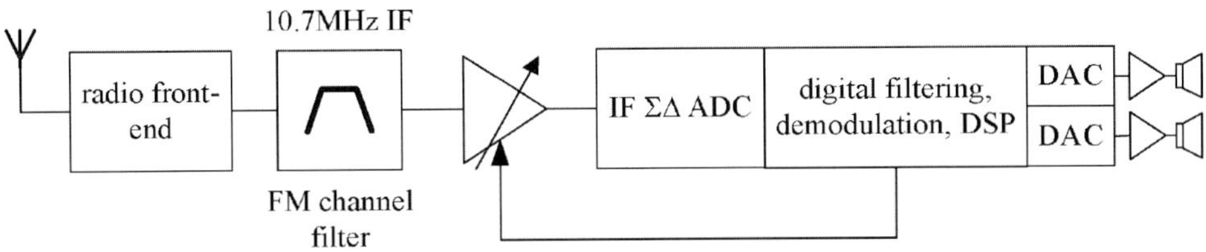

Figure 8.2: *FM/AM radio with digital demodulation*

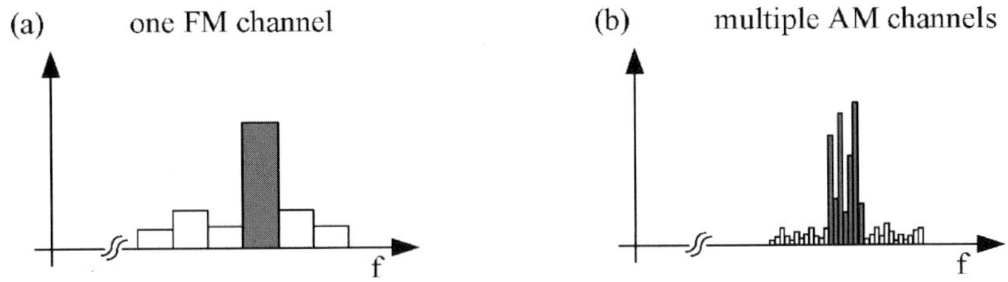

Figure 8.3: *Input spectrum to VGA and ADC in FM (a) and AM (b) mode*

signal processing is already applied in car radios. Hence, this is a further stimulus to digitize the receiver as well.

8.1.2 Radio with digital demodulation

A FM/AM receiver with a higher level of digitization (a.o. digital demodulation) is shown in fig. 8.2. In this solution, demodulation of the radio signal is performed digitally. *An FM SAW channel filter,* usually about 200kHz wide and with ~40dB of adjacent channel suppression, *is still required* to protect the VGA and the ADC from strong interferer channels. *Its specification is however relaxed* compared to that in fig. 8.1 because it doesn't have to provide the full channel selectivity: final channel filtering takes place in the digital domain. The A/D conversion happens at the 10.7MHz IF, both in FM and in AM mode. In fact, it is detailed later on that this IF-ADC consists of a low-pass ΣΔ ADC with a passive mixer integrated in its input stage [38].

Highly-analog signal conditioning in FM mode: In FM mode a single 200kHz channel is converted to digital. The neighboring interferer channels are not completely removed but they are sufficiently suppressed in order not to influence the ADC design (see fig. 8.3.a). Hence, in FM mode, the signal conditioning is a compromise between that of the full-analog channel of section 5.1 and that of the channel in section 6.2 with signal-conditioning in the decimation filter.

Highly-digital signal conditioning and multi-channel A/D conversion in AM mode: Compared to the analog receiver of fig 8.1, the external AM channel filter can be omitted. This reduces the required PCB area, the number of pins, power consumption in I/O

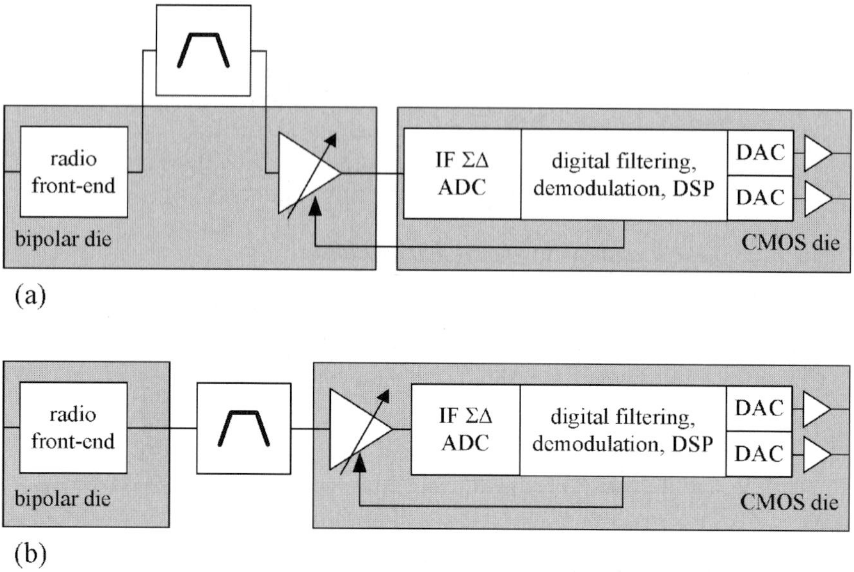

Figure 8.4: *Technology partitioning with bipolar (a) and with CMOS (b) VGA*

drivers, interference, etc. AM channel selection is shifted to the digital domain completely: since the AM channels are only 9kHz wide, about 20 channels pass through the 200kHz FM filter (see fig. 8.3.b)). Hence, in AM mode, this is a highly-digital conditioning channel with multi-channel A/D conversion just as in the channel of section 5.2.

8.2 VGA design

The main research focus of the car radio concept of fig. 8.2 is in the design of the ADC. Still, here, the analog conditioning in the VGA preceding the ADC is presented as well. The original technology partitioning is depicted in fig. 8.4.a: the RF front-end and the VGA are on a bipolar die, the IF-ADC is integrated with the digital processing and the other mixed-signal circuits in a $0.25\mu m$-CMOS technology. In an attempt to alleviate the design of the bipolar VGA, the design of a CMOS VGA is explored (fig. 8.4.b). *The key challenge of the design is in a very high linearity* (i.e. $IM_3 < -80dB_c$ in AM mode) at a differential output swing of $1.4V_{pp}$ at a supply of only 2.5V.

Integrating the VGA in CMOS has obvious benefits with respect to pin-count, reliability and compatibility as the gain control loop can then be closed on one die. The fixed pre-amplification is kept on the bipolar IC in order to maintain noise immunity at the interface between both ICs. The interface between the ICs is differential. The system furthermore requires four gain steps of about 6dB each and a differential output swing of $1.4V_{pp}$ from the VGA to the 62.5kOhm input impedance of the ΣΔ ADC (see section 8.3).

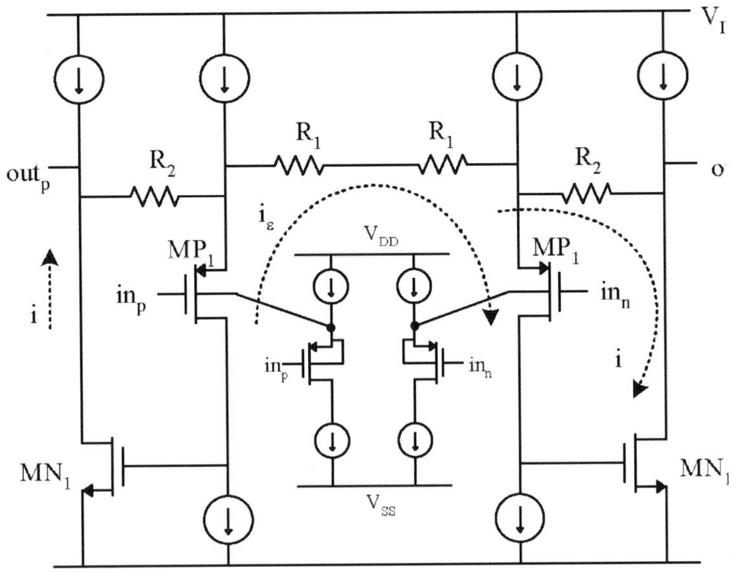

Figure 8.5: *Highly linear VGA with feed back*

8.2.1 Highly linear VGA design

A highly-linear VGA is depicted in fig. 8.5. Its basic operation and an AC analysis are discussed next.

Basic operation

Fig. 8.5 depicts a two-stage amplifier with negative feedback where the input signal is applied to the positive input terminal [92]. The current sources force the input transistor MP_1 to conduct a fixed quiescent current. Thus MP_1 operates as a source follower and copies the input signal at its gate over resistance R_1, being a highly linear conversion impedance. The second amplifying stage acts as a class-A amplifier. Transistor MN_1 provides the signal current. It may be very non-linear, as its distortion is reduced by the gain in the preceding stage. Because of the feedback, transistor MP_1 only needs to be linear with respect to a small error current i_ϵ instead of the entire signal current i. The larger the loop gain at the IF frequency, the smaller the error current and the less linearity is required of MP_1. Thus, *the presented VGA topology achieves low distortion by exploiting the gain-bandwidth product available in advanced CMOS-technologies instead of the (decreasing) inherent linearity of the devices.*

P-MOS devices have been used for the input pair of the amplifier. This allows controlling the back-gate voltage: a bias circuit for the Nwell of the input transistors prevents back-gate modulation. At the same time, this prevents bandwidth limitations or linearity problems due to the Nwell capacitance. The bias circuit consists of a replica of the input branches that copies the AC input signal to the Nwell of the input P-MOS transistors. By consequence, the charging current of these nodes does not interfere in the signal path.

The input common-mode voltage is controlled by a feed back loop that is outside the present scope. The output common-mode voltage is set by the ΣΔ ADC that follows this VGA. All four branches conduct $500\mu A$. This quiescent current is needed to meet the noise and bandwidth requirements.

The closed-loop gain of the VGA approximately equals $1+R_2/R_1$. It is shown in eq. 8.4 that this approximation still holds around the 10.7MHz input frequency. The gain variation is implemented by varying R_1. Because of the chosen resistances and including the on-resistance of the switches, gain settings of 5.8dB, 12.5dB and 17.0dB are obtained. (The requirement for a fourth gain setting (see introduction) is met when by-passing the VGA and thus realizing 0dB of gain.) Switching R_1 instead of R_2 has an advantage in terms of loop stability (see below) and for linearity. Details on the latter are described in [93], as well as comments on the accuracy of the gain setting.

Loop analysis

For this analysis, we refer to the universal amplifier model of fig. 8.6.a and to the parameters defined in the single-ended schematic of fig. 8.6.b. The open loop gain A, the feed back factor k and the input factor d of the VGA can be calculated:

$$A = \frac{g_{m1}}{sC_1} \frac{g_{m2}(R_2 + R_1//\frac{1}{g_{m1}})}{1 + sC_2(R_2 + R_1//\frac{1}{g_{m1}})} \tag{8.1}$$

$$k = \frac{R_1//\frac{1}{g_{m1}}}{R_2 + R_1//\frac{1}{g_{m1}}} \tag{8.2}$$

$$d = \frac{\frac{1}{g_{m1}}}{\frac{1}{g_{m1}} + R_1//R_2} \tag{8.3}$$

The loop gain equals Ak and the stability can be analyzed by solving $1+Ak=0$. The poles and zeroes of the circuit are indicated in fig. 8.6.b. In order to simplify the expressions, the high-frequency pole-zero pair at node *N3*, is left out of the above equations.

In fact, fig. 8.6 shows a simplified schematic. In the actual circuit, Miller compensation is applied to split the poles at nodes *N1* and *N2* in fig. 8.6.b and to guarantee stability. In fact, the stability is conditional since the second pole depends on the load impedance. In the present application, though, the load of the VGA is set by the input impedance of the ΣΔ ADC that is discussed in section 8.3. It represents a resistive load of $62.5k\Omega$. This impedance can be neglected compared to the load of the feed back path.

Finally, it is mentioned that the gain-bandwidth product of the loop exceeds 30GHz in all modes. This guarantees fast settling in case of a transition to a different gain setting.

Assuming strong degeneration (i.e. $g_{m1}R_1 \gg 1$), the closed-loop gain can be ex-

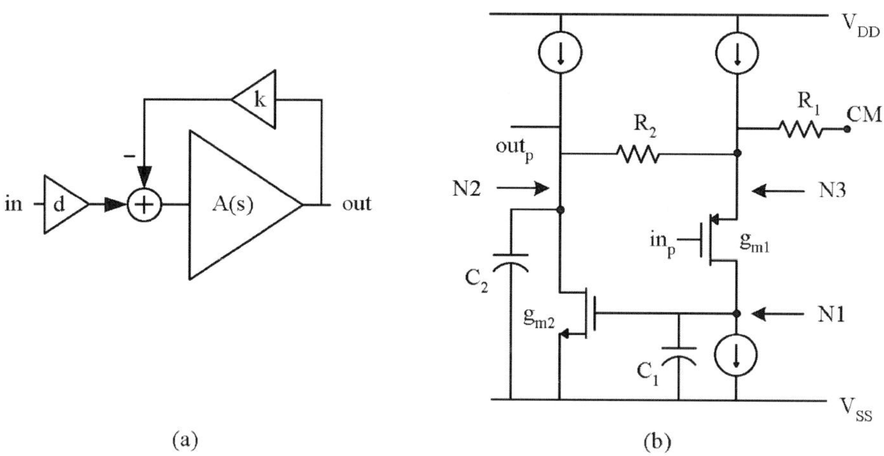

Figure 8.6: *Universal amplifier model (a) and single-ended VGA schematic (b) for analyzing the loop*

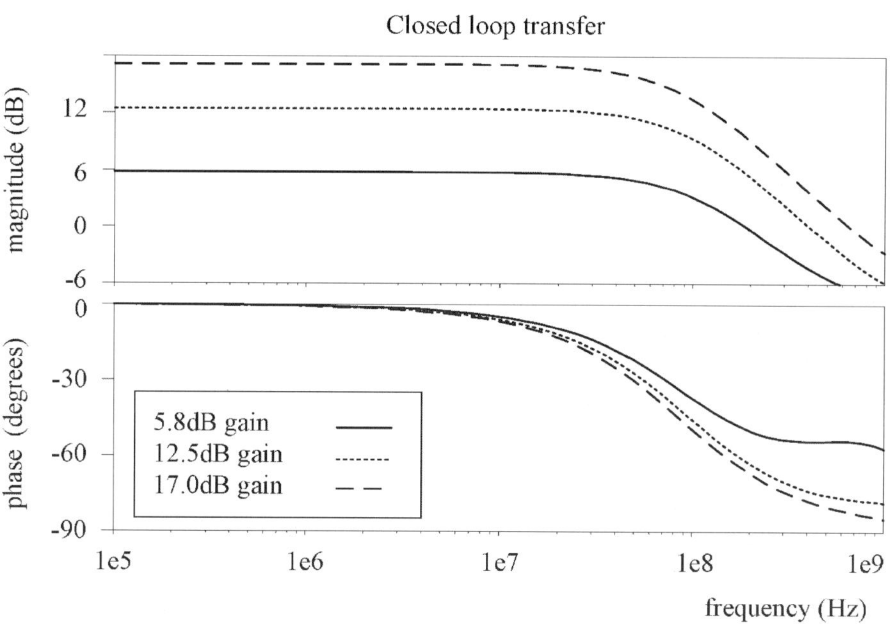

Figure 8.7: *Closed-loop transfer of the VGA (simulated)*

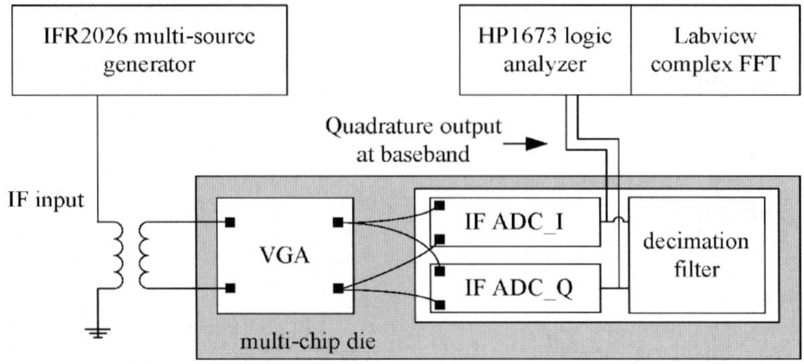

Figure 8.8: *Measurement set-up*

pressed as:

$$\frac{dA}{1+kA} = \left(1+\frac{R_2}{R_1}\right)\frac{g_{m2}}{g_{m2}+sC_1}\frac{1}{1+sR_2C_2} \tag{8.4}$$

Its magnitude and phase are plotted in fig. 8.7 for the three gain settings of the VGA. Hence, the closed-loop gain of the VGA approximately equals $1+R_2/R_1$ for frequencies within the closed-loop bandwidth determined by $f_{-3dB}=g_{m2}/C_1$. In a first-order approximation, the closed-loop bandwidth is the same for all gain settings ($f_{-3dB}\sim 100MHz$) since it is independent of R_1. Similarly, the loop stability is independent of R_1 as well. On the contrary, any gain setting approach varying R_2 would interfere with loop stability: i.e. it would shift pole at node $N2$ in fig. 8.6.b.

8.2.2 Evaluation

The high linearity target requires a dedicated measurement set-up to evaluate the VGA. This set-up is discussed first. Next, the measurement results are listed and evaluated in view of the radio requirements, of the simulation results and of the power/performance relations derived in chapter 4. Finally, the overall VGA performance is bench-marked with state-of-the-art published designs.

Measurement set-up

The VGA has been processed as a stand-alone circuit with an active area of $0.038mm^2$ in $0.25\mu m$ CMOS technology. The measurement set-up is drawn in fig. 8.8. A highly-linear, multi-source generator allows two-tone intermodulation tests around the 10.7MHz IF frequency. Using a transformer, a differential input signal is applied to the VGA.

The output of the VGA is measured using the IF-ADC that is discussed in the next section. This ADC has integrated quadrature mixing and, by consequence, provides a complex output signal around a low-IF frequency. It was fully characterized before-hand. The VGA and the ADC are on separate dies but have been packaged together in a multi-die module (see fig. 8.9). The bitstream output of the ADC is captured by a logic ana-

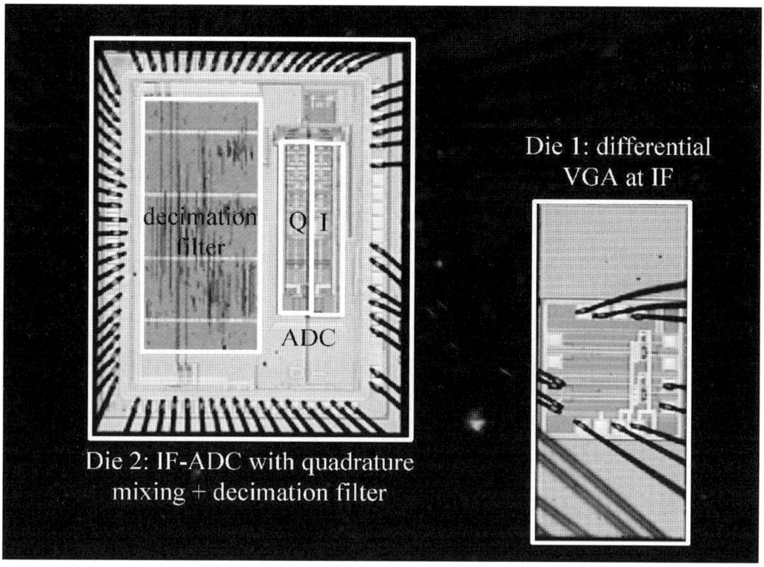

Figure 8.9: *Micrograph of multi-die module combining VGA (die1) and IF-ADC (die2)*

lyzer. Further analysis can then be conducted in the digital domain and at the baseband frequency. This set-up is chosen because of the stringent linearity requirements.

Measurement results

Fig. 8.10 shows the output spectrum in case of a two-tone measurement at an intermediate gain of 12.5dB. Two tones -equal in amplitude- are applied such that they add up to a $1.4V_{pp}$ output signal for the VGA. Since this amplitude corresponds to the full-scale input level of the ADC, each tone appears at $-9dB_{FS}$ at the digital output. The 3rd order intermodulation components fall into the band of interest and are at a level of $-71dB_c$, i.e. 71dB below the carrier tones. From the evaluation of the ADCs, it can be concluded that the measured IM_3-distortion is dominated by the VGA non-linearity. The other spectral tones (image frequencies and DC offset) are generated by the IF-ADC with quadrature mixing.

The noise floor in fig. 8.10 combines the contributions of the VGA and the ADC. VGA and ADC separately, have an output-referred *DR* of 96 and 97dB respectively (measured in 9kHz bandwidth for an AM channel). Taking into account the gain range of the VGA the entire CMOS IF receive path achieves an input-referred *DR* of 110.5dB in a 9kHz AM channel. The output-referred noise density of the VGA is 84nVrms/sqrt(Hz). It is dominated by the noise of the current sources and, thus, is largely independent of the gain settings. The noise density rises for frequencies larger than 300kHz due to the noise shaping of the ΣΔ ADC.

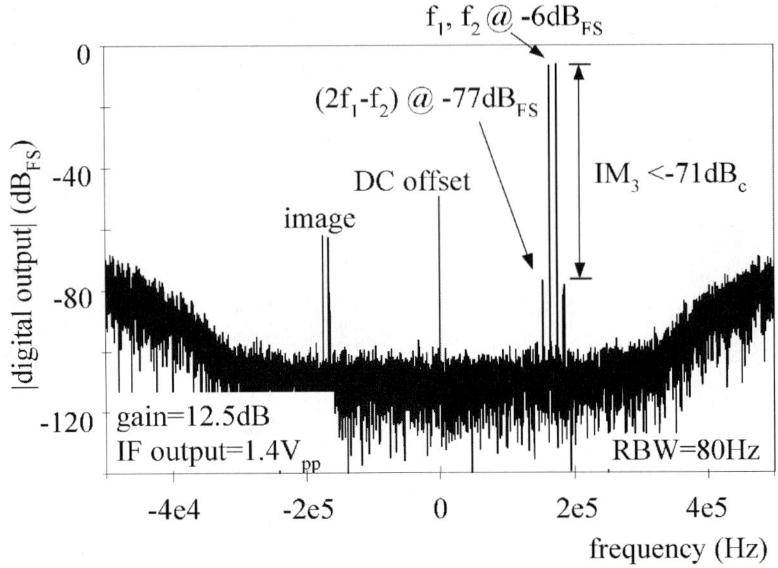

Figure 8.10: *Output spectrum (after quadrature mixing nd ΣΔ A/D conversion) for an IM_3 test*

Table 8.1 summarizes the major performance parameters of the VGA. Because the VGA architecture is based on feedback its performance is robust with respect to variations in supply voltage, quiescent current, processing, etc. Moreover, it is very well scalable to new CMOS technologies.

Discussion

The presented CMOS VGA provides a large output signal swing at IF while keeping IM_3-distortion low (see benchmark in the following section). Still, the linearity is not good enough for the AM radio application because of the multi-channel aspect. A potential solution is to scale down the signal swing in the channel: IM_3-distortion then decreases quadratically. Alternatively, the bipolar VGA (at a higher supply voltage) is used or the VGA can be integrated into the ΣΔ ADC. The latter solution is discussed for the ΣΔ ADCs targeting Bluetooth specifications (sections 9.3 and 9.3).

The measured distortion is higher than what is predicted from simulations. This can be due to a.o. imperfections in the device modeling, numerical problems in the distortion simulation or artifacts in the circuit modeling or the measurement set-up. The discrepancy was investigated in various tests.

- *Effect of decreasing the signal swing:* The IM_3-distortion drops quadratically. Hence, the circuit is not in compression and the distortion is due to a weak non-linearity.
- *Effect of varying the supply voltage:* The IM_3-distortion is largely independent of the supply voltage.

Table 8.1: *Measured VGA performance*

differential output range	$1.4V_{pp}$
gain settings	$5.8dB/\,12.5dB/\,17.0dB$
IM_3 [a]	$<-72dB_c/\ <-71dB_c/\ <-67dB_c$
output noise density	$<84nV_{rms}/\sqrt{Hz}$
$-3dB$-bandwidth	100MHz
$I_{quiescent}$ (VGA + bias circuit)	(2.1+0.6)mA @2.5V
active area	$0.038mm^2$
technology	$0.25\mu m$-CMOS, 1P, 6Al

[a] at 10.7MHz and $1.4V_{pp}$ output

- *Effect of varying the quiescent current:* In the highest gain settings, the IM_3-distortion increases slightly if the quiescent current is decreased below its nominal value. For the setting with gain 5.8dB, the IM_3-distortion is independent of the quiescent current. This is also the case -for all gain settings- if the quiescent current is increased beyond the nominal value.

The latter experiment is illustrated in the graph of fig. 8.11. In the setting with 17.0dB or 12.5dB of gain and with a quiescent current below the nominal current, the IM_3-distortion seems to follow a slope of -1 or $-2/3$ as a function of the quiescent current. This relation is expected for the non-linearity of a transconductor or a degenerated transconductor (eq. 4.7), respectively. In this bias region, and for these gain settings the input transconductor of *MN1* is only weakly degenerated. Hence, it may cause the dominant distortion here.

Beyond the nominal quiescent current or for a gain of 5.8dB, the distortion slowly rises. However, the distortion remains low over a large variation in the quiescent current. Therefore, it may be due to a dominant non-linearity outside of the actual VGA loop.

The curves in fig. 8.11 also prove that *the VGA is biased in a region with an optimal power/performance ratio; i.e. this is a local minimum in terms of distortion.*

Benchmark

The presented VGA does not meet the very challenging IM_3 requirements set by the multi-channel aspect in AM mode. Still, compared to published CMOS VGAs, it performs very well. This demonstrated by the comparison to state-of-the-art published designs with similar bandwidth and gain targets in table 8.2. All these designs use local or global feed back. By consequence, the power/performance relation of eq. 4.7 with $\alpha = 2/3$ is valid (assuming a transconductor limits the performance) and the performance of these designs

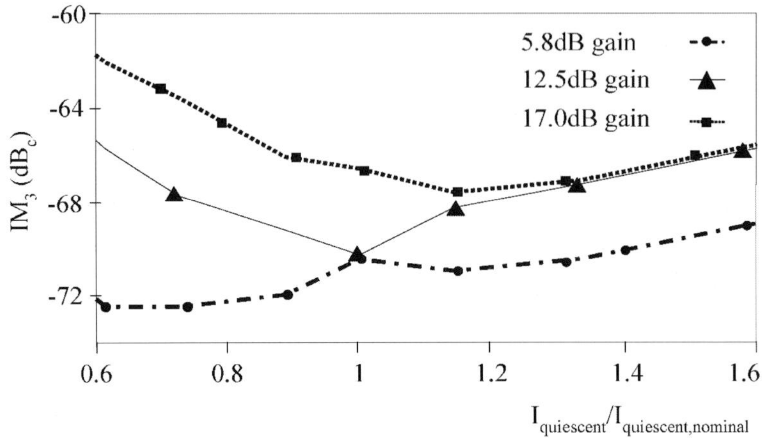

Figure 8.11: *IM_3-distortion as a function of the quiescent current (normalized to the nominal quiescent current)*

can be compared using the FOM of eq. 4.15. Although this FOM has been derived for a ΣΔ ADC, it is equally applicable to analog circuits using feedback (see discussion on page 52).

This comparison proves the excellent power/performance of the presented VGA. [95] achieves even lower distortion but benefits from a higher supply voltage (reducing a.o. the non-linearity of the switches).

8.3 ADC design

Depending on the gain range of the VGA, the *DR* at the input of the ADC is about 70dB over 200kHz bandwidth in FM mode. In AM mode, again depending on the VGA range, the required *DR* at the ADC input may be larger than 90dB in 9kHz. This means that in AM mode a lower noise density must be achieved than in FM mode. The multi-channel aspect also results in severe requirements in terms of linearity and spurious-free *DR*: distortion may introduce intermodulation or other spurious components into the wanted channel. This makes the FM/AM ADC more difficult to realize in AM than in FM mode.

8.3.1 Conventional solutions

Digitization in fig. 8.2 may be performed by a wide-band ADC [90]. A resolution of 10 bits in 11MHz bandwidth may be sufficient to achieve the noise specifications. However, the linearity requirements for AM are difficult to meet at the 10.7MHz input frequency. Therefore, an analog AM channel filter is still necessary, which is applied after mixing to a second IF of 450kHz.

Table 8.2: *Comparison of VGA performance*

Reference	[94]	[95]	This work [93]
gain range	$-2 \sim 12$dB	$0 \sim 19$dB	$5.6 \sim 17.0$dB
f_{-3dB}	15MHz	125MHz	100MHz
output swing	$1V_{pp}$	$2V_{pp}$	$1.4V_{pp}$
IM_3 [a]	$-56.5dB_c$	$-74dB_c$	$-67dB_c$
output noise density	$67nV_{rms}/\sqrt{Hz}$	$55nV_{rms}/\sqrt{Hz}$	$84nV_{rms}/\sqrt{Hz}$
current consumption	4mA	6.4mA	2.1mA
supply voltage	5V	3.3V	2.5V
technology	$0.5\mu m$-CMOS	$0.35\mu m$-CMOS	$0.25\mu m$-CMOS
FOM of eq. 4.15	$1 \times 10^{-18} J$	$1 \times 10^{-19} J$	$2 \times 10^{-19} J$

[a] for inputs around 10MHz and at maximum gain

Alternatively, bandpass ΣΔ modulation may be used [96], [97], [98], [99], [100], [101] and [102], providing the required resolution in the bandwidth of interest only. Solutions for AM [97] and for FM [98], [99] have been shown but with limited performance and relatively high power consumption. Further development resulted in combined solutions for AM and FM [96], [100]. Again, the input stage has to be very linear at the 10.7MHz IF to prevent intermodulation of neighboring channels in AM mode. This is very difficult.

A different approach is to use a ΣΔ phase-locked loop (PLL) to digitize and demodulate the FM radio signal directly after the limiter in fig.2 of [103].

Because of the high DR of the radio signal and the intermodulation distortion requirements, none of these techniques eliminates the need for a high-Q channel filter.

8.3.2 ΣΔ ADC with integrated passive mixer

The ADC that is presented here mixes the 10.7MHz input signal to a second low intermediate frequency, and uses low-pass, continuous-time ΣΔ modulation to digitize the signal. To this purpose, a passive mixer is integrated into the input stage. This technique was first presented in [38]. Here, it is briefly reviewed together with the global frequency planning of the IC. First, the choice for a low-IF topology is motivated and the consequent requirements on image rejection are discussed.

Low-IF topology: The second mixer may convert the wanted channel to DC ("zero-IF") or to an offset frequency ("low-IF" or "near-zero IF"). Both methods have their advantages and disadvantages [104]. Using zero-IF, the required ADC bandwidth is as low as possible. The main disadvantage is the sensitivity to flicker noise and DC offset. The latter may be introduced by circuit mismatch, oscillator self-mixing or second-order distortion in the mixer. Flicker noise as well as DC offset are located in the middle of the wanted channel and therefore cannot be removed without loss of part of the signal. This is the main reason why mixing to a low IF is preferred in this application: the low IF frequency is chosen such that flicker noise and DC offset are outside of the signal bandwidth. A quadrature signal path is now needed for suppression of the image channels. Still, this topology remains efficient in terms of power consumption.

Image rejection: Gain and phase mismatch in the quadrature paths leads to leakage of signals around the image frequency into the wanted channel. Using zero-IF the signal at the image frequency is the mirror of the wanted channel, and therefore the requirements for image rejection are relatively relaxed in that case. However, when a low IF is used, the signal at the image frequency of the wanted channel is its adjacent channel. Since the adjacent channel may be significantly stronger than the wanted channel, the image rejection requirements are much more severe for a low-IF than for a zero- IF topology [104]. This implies that accurate gain matching and a precise 90 degrees phase difference between the I and the Q paths is required. The image rejection (IR) is defined as:

$$IR = \frac{\text{sensitivity to the wanted channel}}{\text{sensitivity to the image channel}} \tag{8.5}$$

and can approximately be expressed as:

$$IR = 10 \log \frac{4}{\left(\frac{\Delta A}{A}\right)^2 + (\Delta\phi)^2} \tag{8.6}$$

where $\Delta A/A$ represents the relative gain mismatch between I and Q, and $\Delta\phi$ is the phase error in radians [105]. The FM/AM application requires about 80dB of image rejection. The channel filter at IF (fig. 8.2) attenuates the adjacent channel (in FM mode) by about 40dB, so another 40dB of rejection must be provided by the matching of the I and the Q path. Assuming equal contributions from gain and phase error, a gain error below 1.4% is required, and the phase error must be less than 0.8degrees.

Integrated mixer and clocking scheme: A passive mixer can easily be integrated in the input stage of a continuous-time $\Sigma\Delta$ ADC. In [38], it is shown that this technique enables a high- resolution, high-linearity IF-ADC with low power consumption. Fig 8.12 shows how this technique is applied to the ADC for the FM/AM application. The input signal around the 10.7MHz IF is applied as a differential voltage. It is converted to currents by input resistors R_{in1} to R_{in4}. Passive mixing is implemented by the switches at the inputs of the $\Sigma\Delta$ ADC, which are low-ohmic (virtual ground) in this case. The on-resistance of the switches is much smaller than the input resistance. Thus, high linearity is achieved because the resistance of the switches is only weakly modulated. The power consumption of the passive mixer is negligible.

The low-IF topology requires quadrature mixing and two parallel ADCs for the I and the Q path. The combined output of both ADCs represents a complex signal that is

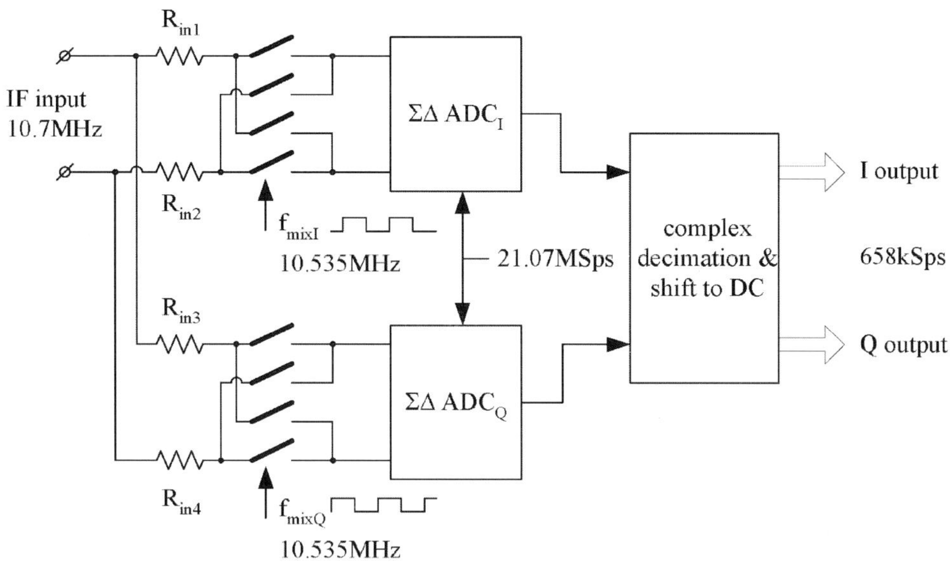

Figure 8.12: *Integration of a passive mixer into the FM/AM ADC and clocking scheme*

decimated in the digital, complex domain. In addition, the digital filter also performs the final frequency shift to DC [106].

Fig. 8.12 also shows the clocking scheme for the ADC. The master clock at a frequency of 42.14MHz is generated by an on-chip crystal oscillator. From this master clock, the 21.07MHz sampling frequency for the ΣΔ ADC and the 10.535MHz frequency for the mixer are derived. The latter frequency is provided twice, with a 90 degrees phase difference between the clock for the I and for the Q path. After the mixer, at the input of the ADC, the signal is centered on the difference frequency of 165kHz. In the ΣΔ ADC, the signal is 32 times over-sampled. At the output of the decimation filter a Nyquist sample rate of 658kHz results.

Architecture

The ΣΔ ADC is based on the topology of fig. 3.4 and discussed in [7] and in section 3.2.2. Here, the modulator is 32 times over-sampled and uses return-to-zero pulses in the feed back in order to minimize inter-symbol interference. A 5th-order loop filter with complex conjugated poles (i.e. local feedback loops with coefficients b_3 and b_5) is implemented (see fig. 8.13). These poles provide additional filter gain within the signal bandwidth and appear as notches in the shaped quantization noise spectrum. One of the notches is located at the 165kHz IF frequency, the other one at the edge of the signal band. The loop filter is implemented by means of simple transconductor-C integrators with feed forward coefficients a_i. These feed forward coefficients provide first-order roll-off at unity open-loop gain for stability reasons. Large signal stability is guaranteed by clipping the integrator

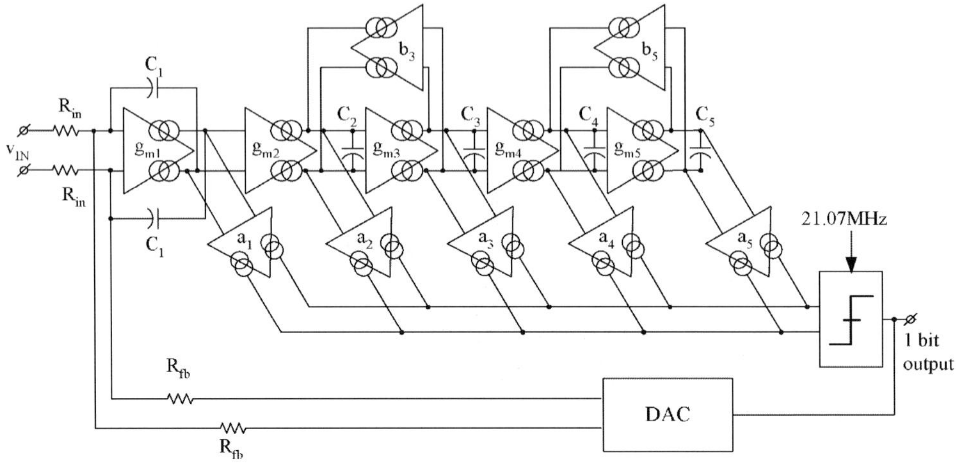

Figure 8.13: 5th*-order* $\Sigma\Delta$ *ADC*

outputs, starting at the fifth integrator (graceful degradation). The resonators introducing the complex poles are implemented using local feedback transconductors $g_{m,b1}$ and $g_{m,b2}$. The resulting shaped quantization noise spectrum, assuming ideal circuit elements, is shown in fig. 8.14.

In practice, various non-idealities affect the output spectrum. These are due to a.o. the finite quality factor of the resonators, component spread and additional noise sources.

Quality factor of resonators: Transconductors g_{m2}, g_{m3} and b_3 in fig. 8.13 constitute the first resonator (located at the edge of the signal bandwidth). Taking into account the output impedance of these transconductors, the resonance appears at:

$$\omega_r = \sqrt{\frac{1 + b_3 R_2 g_{m3} R_3}{R_2 C_2 R_3 C_3}} \tag{8.7}$$

where R_2 and R_3 represent the parasitic impedance that is in parallel with the integration capacitors C_2 and C_3, respectively. In case of a high-impedance node, eq. 8.7 is simplified to:

$$\omega_r \approx \sqrt{\frac{b_3 g_{m3}}{C_2 C_3}} \tag{8.8}$$

In practice, the presence of R_2 and R_3 causes a shift of the resonance frequency to a lower value.

Likewise, these impedances reduce the quality factor Q of the resonance to:

$$Q = \frac{\omega_r}{\frac{1}{R_2 C_2} + \frac{1}{R_3 C_3}} \tag{8.9}$$

By consequence, the notches in the noise shaping degrade.

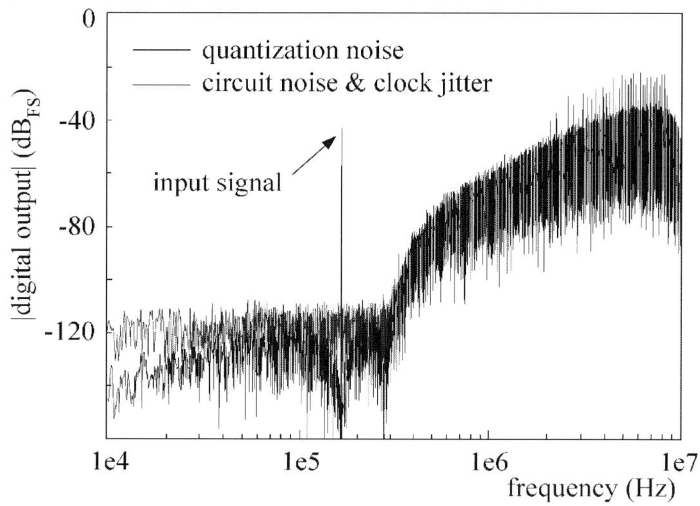

Figure 8.14: *Simulated output spectrum with quantization noise, circuit noise and jitter induced noise*

Component spread: Component spread due to a variation in processing, temperature, supply voltage, etc., causes the noise shaping curve of fig. 8.14 to shift to the left or the right. Table 8.3 lists the expected variation on the degeneration resistance and the integration capacitance determining the time-constants of the loop filter for the $0.25\mu m$-CMOS technology the ADC is implemented in. From this, a 4σ-deviation of less than 25% is expected. Since the FM channel bandwidth typically equals 200kHz and the nominal conversion bandwidth of the ADC is $\sim 330kHz$, a worst-case negative spread of the noise shaping in combination with a high operation temperature, can be accommodated.

Additional noise sources: In fig. 8.14, additional, white, noise sources are added to the ideal quantization noise spectrum. The dominant noise source is due to the thermal noise of the input and the feedback DAC resistors and the input transistors. (Each input resistance equals $62.5k\Omega$, each DAC resistance equals $100k\Omega$.) As discussed in chapter 4, this noise source can only be reduced at the expense of an increase in quiescent current and therefore it is left to be dominant. The corner frequency of the flicker noise is outside the bandwidth of interest.

Another limiting factor may be clock jitter. The approximation in [7] and simulations show that the rms-value of the clock jitter must be below 7ps in order not to contribute to the overall *DR*. This accuracy can be achieved with an on-chip crystal oscillator.

Circuits

The input stage of the ADC in fig. 8.13 is implemented as an active RC-integrator (fig. 8.15). It is based on a simple single-stage OTA, with wide input transistors operating

Table 8.3: *Variation on component values in the $0.25\mu m$-CMOS technology*

Processing spread		Temperature [5; 125] deg C	
P^+-poly resistance	gate ox. capacitance	P^+-poly resistance	gate ox. capacitance
$4\sigma = 10\%$	$4\sigma = 6\%$	$[-0.2\%; +10\%]$	$< 1\%$

in weak inversion with $g_m = 4.3mA/V$. It provides the virtual ground node where the mixer can be conveniently implemented using N-MOS switches and where the DAC feedback current is subtracted from the input signal. Poly-silicon resistors and gate-oxide capacitors are used. Nulling resistors in series with the integration capacitors compensate for the right-half plane pole that is introduced by this structure.

The P-MOS tail current source is degenerated with transistors in the triode region. These perform the output common-mode control since they match with a replica circuit that is biased at the target voltage.

The current consumption of the input stage, which is determined mainly by the noise and the distortion requirements (see section 4.4), is $500\mu A$. Notice, P-MOS devices are used in order to prevent back-gate modulation and to provide isolation from the low-resistivity substrate.

The resonator stages consist of two $g_m C$-integrators and a feedback transconductor. The implementation is shown in fig. 8.16. All, use a degenerated, differential pair input. The current consumption of a complete resonator is $240\mu A$. The feed forward coefficients are implemented as degenerated differential pairs as well. Their outputs are all fed to the current input of the quantizer. It consists of a cross-coupled latch [7].

Decimation filter

The block diagram of the integrated, complex decimation filter is shown in fig. 8.17. The 21.07MHz 1-bit output of the I and the Q $\Sigma\Delta$ ADC is first filtered by a CIC filter and decimated by a factor of 16. Then, the signal band of interest, around 165kHz, is shifted to DC by a complex mixer (CORDIC algorithm, [106]) and, simultaneously, decimated by a factor of 2. Next, a final decimation by 2 is performed to an output sample rate of 330kHz and a word-length of 16 bits.

Finally, it should be noted that careful layout of the DAC, the mixer and the input stage is indispensable. Crosstalk from the large amount of quantization noise around half

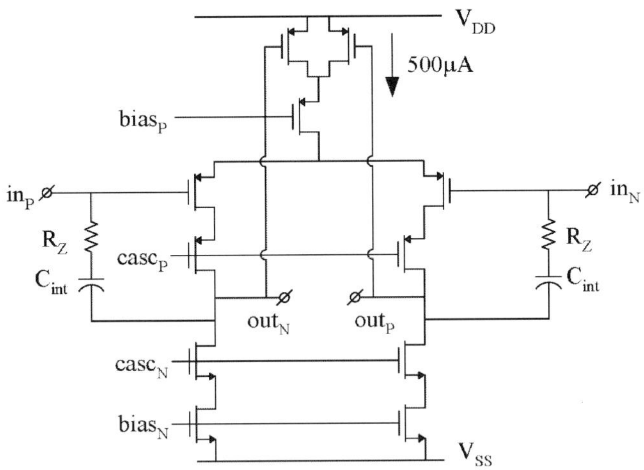

Figure 8.15: *Input stage of ΣΔ ADC*

the sampling frequency towards the IF input must be minimized. Layout symmetry is essential to reduce gain and phase mismatch between the quadrature paths for best image rejection.

8.3.3 Evaluation

A prototype IC was originally fabricated in a $0.35\mu m$ digital CMOS technology. It includes an I and a Q ΣΔ ADC with IF mixers, the complex decimation filters and the digital shift to DC. Furthermore, a bandgap reference and a crystal oscillator are integrated.

The IC is successfully scaled to a $0.25\mu m$-CMOS technology. This resulted in lower power consumption and a smaller chip area of $1.35mm^2$. The digital filters measure $0.75mm^2$ and the analog part, including the bandgap is $0.55mm^2$. The oscillator is $0.04mm^2$. The ADCs and the decimation filter are indicated on the micrograph of fig. 8.9.

Measurement set-up

The measurement results of the next silicon in the $0.25\mu m$-CMOS technology are reported below. Both the 16-bit outputs of the digital block, at baseband, and the bitstream outputs of the I and the Q ΣΔ ADC, at the low IF of 165kHz, are available for evaluation. Differential, current-mode output buffers are used for the bitstream outputs in order to minimize crosstalk. Below, though, the output spectra at baseband are shown.

The set-up is similar to that for the evaluation of the VGA in fig. 8.8. Of course, the VGA is left out of the set-up. A HP8640B signal generator is used for low-noise single-tone measurements and the IFR2026 generator is used for multi-tone measurements requiring high linearity.

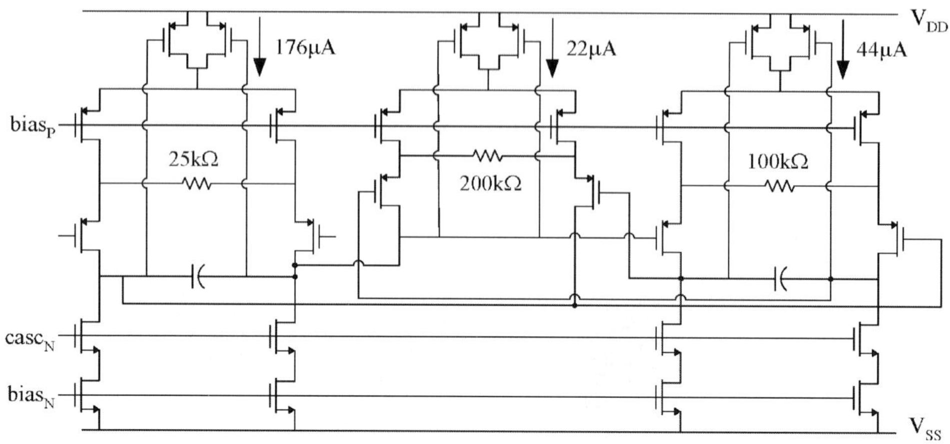

Figure 8.16: *Resonator implementation*

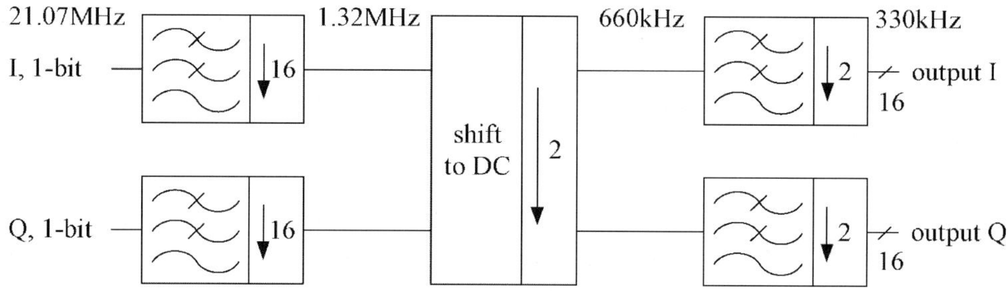

Figure 8.17: *Block diagram of complex decimation filter*

Measurement results

Fig 8.18 shows a complex output spectrum where the wanted signal is near the 165kHz low-IF frequency. It is measured at the bitstream output of the ΣΔ ADCs. The maximum, differential input signal of the ADC is $1.4V_{pp}$ (for each ADC). Measured *DR* is 97dB in 9kHz (AM) and 82dB in 200kHz (FM) bandwidths. As expected, thermal noise dominates in the bandwidth of interest.

Especially in AM mode, the peak-SNR is important, since a rise of the noise floor caused by a large neighboring channel, decreases the sensitivity for the wanted channel. Measured peak-SNR is 94dB in 9kHz at the maximum input signal of $1.4V_{pp}$. The 3dB increase of the in-band noise at maximum input is probably caused by distortion of the high-frequency quantization noise in the loop filter of the ΣΔ ADC.

Intermodulation distortion is especially important in AM mode, since the input signal contains multiple AM channels. Distortion of strong unwanted channels may deteriorate

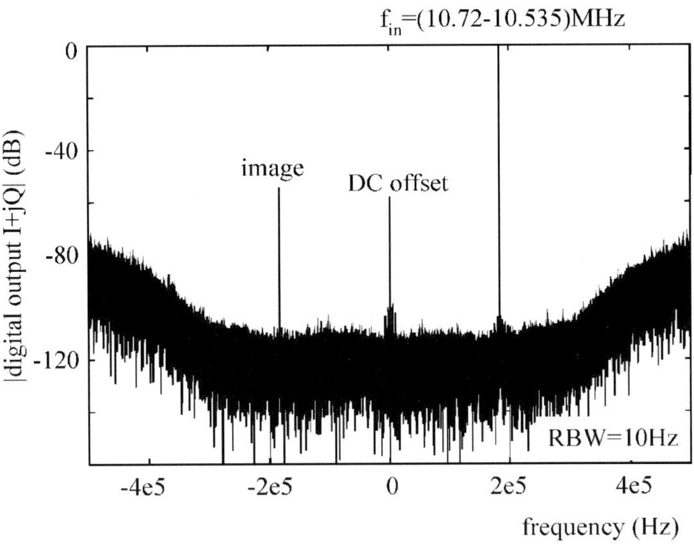

Figure 8.18: *Measured complex output spectrum*

reception of the wanted channel. IM_3 was measured by applying a full-scale, IF input signal consisting of two $0.7V_{pp}$ tones at 10.710 and 10.715MHz, so that the ADC input signal is maximal. The resulting IM_3 is $-84dB_c$ (i.e. the intermodulation components are 84dB lower than the wanted signal) and decreases rapidly for small signals.

Non-linearity also leads to cross-modulation. This effect is well-known in AM receivers: when listening to a wanted channel that is not, or weakly, modulated, the modulation of a strong unwanted channel may be heard, and thus disturbs the reception of the wanted channel. Although this is an important test for AM-receivers, it is outside of the present scope. For details, the reader is referred to [2]. The same is true for the evaluation of image rejection. Finally, offset is measured. The test ICs show offset with $\sigma = 0.7mV$. In the present application, this offset is outside the band of interest because of the low-IF topology.

The total power consumption of the prototype IC is 36mW from a 2.5V supply voltage. Each ΣΔ ADC consumes 4mW and the complex digital decimation filter requires 11mW. The crystal oscillator consumes another 1.6mW. A major power consumption is due to the current-mode, bitstream output buffers. These are for evaluation purposes only and do not contribute during normal operation. Table 8.4 summarizes the main performance parameters of the IF-ADC. Notice, the overall performance of the complex configuration is listed; i.e. the performance of the I and Q ADC operating together. Notice that both FOM_{SINAD} (eq. 4.4) and the modified, elaborate FOM of eq. 4.15 are listed. The first FOM allows a first-order comparison with published ADCs. The second FOM is used for comparison to the performance of the VGA in the previous section.

Table 8.4: *Measured performance of complex ADC (I and Q together)*

full-scale input	$1.4V_{pp}$ (differential)	
input IF	10.7MHz	
sample rate	21.07MHz	
Nyquist sample rate	660kHz	
	AM (9kHz)	**FM (200kHz)**
DR	97dB	82dB
peak-*SNR*	94dB	79dB
IM_3	$-84dB_c$	
cross-modulation	$-91dB_c$	
IR (20 samples)	$>49dB$	
	ΣΔ ADCs	**Dig. filters**
$I_{quiescent}$	3.2mA @2.5V	4.4mA @2.5V
active area	$0.55mm^2$	$0.75mm^2$
technology	$0.25\mu m$-CMOS, 1P, 6Al	
FOM_{SINAD} (eq. 4.4)	$2.5 \times 10^{-16} J$	
FOM of eq. 4.15	$2 \times 10^{-18} J$	

Discussion

The measurement results are in accordance with simulations and the design meets the target specifications. Hence, the presented ADC enables the radio architecture of fig 8.2. An FM/AM radio with increased flexibility, higher quality and more features results. Moreover, an AM channel-select filter can be saved because of the multi-channel conversion in this mode. This represents a significant cost reduction.

The successful realization of the IF-ADC and the presented results are the basis for further development in this area. The design is further scaled to $0.18\mu m$-CMOS and adapted to encompass the IBOC digital radio standard. This is described in [107].

Benchmark

Table A.1 in appendix A gives an overview of published ΣΔ ADCs. Benchmarking the AM performance is difficult because only few ADCs target a similar bandwidth. On the contrary, many ADCs with a bandwidth of 200kHz have been reported and these are used to benchmark the ADC in FM mode. From this comparison, it is clear that the presented ADC combines a high *DR* and excellent linearity with very low power consumption. This translates into the best FOM reported for a 200kHz-bandwidth ADC in CMOS technology (to the authors knowledge).

Comparing to ADCs with a bandwidth of 270kHz, mostly targeting GSM, the design reported in [84] in $0.18\mu m$-CMOS excels: it achieves a better FOM. That design uses a lower supply voltage and larger input swing of $3V_{pp}$ but these effects have already been taken into account in eq. 4.15. Hence, the better power/performance balance may stem from the larger over-sample ratio (see discussion on the Shannon theorem and power in analog circuits on page 30).

8.4 Evaluation of the channel

The reported results are evaluated in relation to the discussion on power/performance relations in chapter 4. Next, the presented channel for the highly-digitized FM/AM radio is benchmarked with other published solutions.

8.4.1 Discussion

First, the power/performance relation achieved for the analog conditioning part (i.e. the VGA) is compared to that of the ADC. Next, extrapolation of the overall power/performance relation of the channel, in case of further digitization, is discussed.

Analog conditioning versus ΣΔ ADC

The presented VGA, in the highest gain setting, achieves a 10 times lower FOM (according to eq. 4.15) than the ΣΔ ADC. This is mainly due to the fact that it achieves the same *DR* for a smaller input signal (while its power consumption remains slightly lower). The worse power/performance relation of the ADC may be due to various reasons:

- the loop filter consists of several stages requiring a minimum quiescent current for robustness, matching, bandwidth, etc. The power consumed for these purposes, however, hardly improves the power/performance relation in terms of *DR* and IM_3;
- the ADC achieves higher performance, especially in terms of linearity. In general, when the target performance of a circuit increases, more devices, other than the input transconductor, start affecting the power/performance relation.

Still, this remark must be put into perspective: it is demonstrated in section 8.3.3 that this ADC achieves state-of-the-art power/performance.

This comparison primarily indicates that a further digitization step, by reducing the amount of VGA, will result in higher power consumption for the overall channel. The ADC must then achieve even higher resolution for a smaller input signal. Because the worse power/performance relation, the power increase in the ADC is higher than what is gained from omitting the VGA .

Table 8.5: *Benchmark of the channel in FM mode*

	analog filter[108] + low-resolution ADC	PLL [103]	band-pass $\Sigma\Delta$ [102]	this work
DR	61dB	85dB	79dB	82dB
IM_3 [a]	$<-60dB_c$	$<-87dB_c$	$<-67dB_c$	$<-84dB_c$
P	12mW+P(ADC)	35mW	77mW+P(dig.)	19mW

[a] at full-scale input

$\Sigma\Delta$ ADC versus digital conditioning

The power consumption of the digital part of the channel is just slightly higher than that of the $\Sigma\Delta$ ADC. The situation corresponds to the graph in fig. 4.4. The graph indicates that, in case a further increase of resolution were required (for instance for further digitization), the power consumption of the $\Sigma\Delta$ ADC would become dominant while that of the digital part increases much less.

From the discussion in this section, it becomes clear that further, straightforward digitization of the FM/AM channel will result in a power increase for the overall channel. As depicted in the graph of fig. 4.4 the power consumption in the ADC becomes dominant and increases dramatically.

8.4.2 Benchmark

This benchmark focuses on the performance in FM mode because more publications are available than for AM operation. In fact, to the author's knowledge, the presented work is the first solution reporting multi-channel A/D conversion in AM radio receivers.

Benchmark with highly-analog channels

[108], [109], [110] and [111] report integrated, band-pass filters around the 10.7MHz IF. A high-quality integrated filter reduces the required external filtering and relaxes the consequent A/D conversion. [108] achieves the lowest power consumption and, therefore, it is listed in table 8.5. Power-wise (anticipating a moderate power consumption in the "low-resolution" ADC) it may enable a more efficient channel than the presented solution with the IF $\Sigma\Delta$ ADC. Performance-wise there is a significant gap, though: this filter targets application in portable radio and achieves less *DR* and linearity.

[103] presents a $\Sigma\Delta$ PLL performing analog demodulation and digitization of the FM signal. It achieves excellent performance but does not allow future development to encompass AM radio as well.

It is concluded that *the power/performance balance of the continuous-time ΣΔ ADC is competitive or better than that of an analog channel with a cascade of dedicated, analog conditioning blocks.*

Benchmark with highly-digitized channels

[90] reports the use of a wide-band ADC for digitizing the FM/AM channel. The ADC achieves 10 bits of resolution in 11 MHz bandwidth. Power consumption is not mentioned.

[96], [97], [98], [99], [100], [101] and [102] present band-pass ΣΔ converters digitizing the signal at the IF. Among these, [102] achieves the best overall performance. Still, its power consumption is significantly higher than that of the other solutions in table 8.5. Moreover, the power consumption in the decimation filter is not yet included. In addition, linearity is difficult at these frequencies.

In FM mode, the presented channel with passive mixing to a low IF and consequent low-pass, continuous-time ΣΔ A/D conversion achieves a better overall power/performance relation than the other solutions in table 8.5. This ADC especially excels in terms of linearity. Therefore, it is the only design allowing the multi-channel conversion of fig. 8.2 in AM mode. All other solutions in table 8.5 have too much distortion and would not be suitable.

8.5 Conclusions

A dual-mode conditioning channel for an FM/AM radio receiver has been presented. In FM mode, the channel remains highly analog. In AM mode, multi-channel A/D conversion enables highly-digitized signal conditioning with digital channel selection. Two key sub-circuits have been presented.

A CMOS VGA operating at a 10.7MHz IF frequency, achieves $-67dB_c$ IM_3-distortion at $1.4V_{pp}$ output. This is competitive with state-of-the-art designs but does not meet the requirements for the present application in AM mode.

A continuous-time ΣΔ ADC achieves a FOM_{SINAD} of $2.5 \times 10^{-16} J$ (from eq. 4.15) and features a very low IM_3-distortion of $-84dB_c$ for $1.4V_{pp}$ input signal. It consumes only 8mW.

The evaluation of these blocks leads to the conclusions that are listed below.

The power/performance balance of the continuous-time ΣΔ ADC is competitive or better than that of dedicated FM-solutions consisting of a cascade of analog conditioning blocks.

In case of further digitization, the power consumption of the ADC becomes dominant and causes a dramatic increase of the power consumption of the overall channel.

The multi-channel aspect, makes the operation in AM mode more challenging than in FM mode, especially because of the very high linearity requirement.

The presented ADC enables the first solution reported for multi-channel A/D conversion in AM radio receivers. As such, it demonstrates the feasibility of a highly-digitized channel for narrow-band systems, even while the linearity challenge is very demanding.

These conclusions illustrate the theory of chapters 4 and 5.

Chapter 9

Conditioning $\Sigma\Delta$ ADCs for Bluetooth

This chapter presents the application of "conditioning $\Sigma\Delta$ ADCs" in a Bluetooth receiver. Design examples in sections 9.2, 9.3 and 9.4 illustrate the theory of section 6.2 on the feed forward ADC with signal conditioning in the decimation filter, of section 6.3 on the conditioning feedback ADC with restricted filtering and of section 6.4 on the FFB-ADC, respectively. Each of these designs enables a Bluetooth-compliant receiver without the need for analog conditioning in front of the ADC. This is different from other published Bluetooth receivers. It is shown that, integrating explicit signal conditioning into the ADC (as done in the designs of section 9.3 and 9.4) promises a major improvement in the overall power/performance ratio of the receiver. This is true both when comparing to a "conventionally digitized" receiver as well as when comparing to a dedicated analog solution.

9.1 System

Bluetooth [112] is a short-range wireless communication system targeting operation in the 2.4GHz ISM band. In most countries, the channels occupy the band from 2.402 to 2.480GHz. The channel spacing is 1MHz and GFSK modulation is used. The latter implies that all information is contained in the phase of the signal. The Bluetooth standard targets a moderate data-rate (i.e. 1Mbps for the early standard) for use in portable applications as for example headsets, printers, etc. By consequence, cost and power drain are the main challenges on the IC solution.

For cost reasons especially, digitization is welcomed since it enables mass-production in an advanced CMOS technology. The key question, however, is whether a digitized solution is competitive with an analog solution in terms of power consumption.

Another motivation for digitizing the Bluetooth receiver, results from the fact that extended standards like Bluetooth Medium-Rate and High-Rate are being defined in order to provide higher data-rates. In these standards, the channel spacing remains unaltered and,

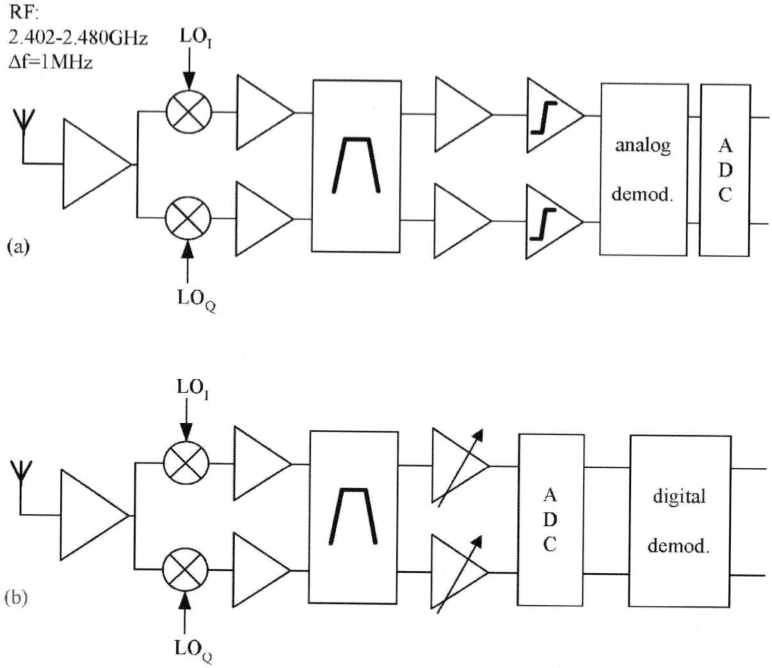

Figure 9.1: *Bluetooth receiver with analog (a) and with digital demodulation (b)*

therefore, a higher-order modulation scheme including amplitude modulation is required to enable this increase. For example, the medium-rate standard targets 2Mbps and uses $\pi/4$-DPSK modulation. This results in a non-constant envelope signal and the simple analog demodulation schemes are not applicable anymore.

9.1.1 Conventional radio with analog demodulation

A conventional Bluetooth receiver with analog demodulation is depicted in fig. 9.1.a. In this application, it is common practice to use a low-IF topology to circumvent the problem of DC errors falling into the signal bandwidth. The baseband part of the receiver consists of a cascade of filter and gain sections. The latter may be clipping to limit the signal to a pre-defined level and can be used to indicate the received signal strength. The bits are recovered by further analog demodulation.

Plenty of low-power, analog demodulation schemes are available for the "simple" GFSK modulation that is used for Bluetooth. They often originate from FM demodulators and allow an easy implementation, resulting in a small area and low cost. For example, a frequency discriminator or a phase-locked loop detector can be used [113]. Alternatively, many zero-crossing detectors have been reported [114], [115], [116] a.o.

Even though these analog demodulators are *robust and simple they require a lot of preceding analog conditioning. More important, the analog receiver architecture is not very future-proof:* adaptation to include amplitude demodulation -required in the extended standards- is not straightforward.

9.1.2 Radio with digital demodulation and analog signal-conditioning

Alternatively, digital demodulation can be implemented [106], [117] as in fig. 9.1.b. Note, the conditioning channel is linear; i.e. the amplitude information is preserved and can be used to improve the GFSK demodulation or to encompass extended standards (with amplitude modulation).

Various Bluetooth receivers with digital demodulation have been reported in a.o. [118], [119], [120] and [121]. They differ in the degree of digitization of the signal-conditioning. For example, [118] uses a 7th-order filter inter-leaved with 84dB of VGA in steps of 1dB, followed by a very simple ADC. Both the filter and the VGA need calibration and tuning to handle spread and drift on the analog parameters.

9.1.3 Radio with digital demodulation, without analog signal-conditioning

A much higher degree of digitization is reported in [121]. The work presented in the remainder of this chapter is part of the overall receiver design described in this reference. The referenced receiver does not need any baseband analog channel filters or VGA. Instead, a high-resolution ADC (see section 6.2) with signal-conditioning in the decimation filter is used. This first solution is discussed in section 9.2. On top of this, in the second solution (section 9.3), some restricted signal-conditioning is integrated into the ADC. In the third solution (section 9.4) the signal-conditioning in the ADC is unrestricted. The other sections in this chapter discuss implementations with more signal conditioning integrated into the ADC.

The single-chip receiver [121] of which the current conditioning channel is a part of, is further detailed in fig. 9.2. Two LNAs are used in parallel to provide sufficient isolation between the I and the Q channels at the mixer input. The LNAs are simple V-to-I converters. The RF output current is down-converted to a low IF frequency of 500kHz by passive mixers, directly driven by a quadrature VCO in a PLL. Hence, the wanted channel extends from 0 to 1MHz at the ADC input. A $\Sigma\Delta$ ADC is used in combination with signal-conditioning in the decimation filter, as described in section 6.2; i.e. the ADC provides high resolution within the bandwidth of the wanted channel while the interferers are present in the noise-shaping part of the output spectrum. The digital block performs channel and decimation filtering as well as demodulation.

The above discussion was generic. From now on, we focus on the radio as introduced in section 9.1.3. More specifically, the signal conditioning at the low-IF frequency is considered. Obviously, this receiver achieves a higher degree of digitization than that of fig. 9.1.b. Because of the absence of analog channel filters and VGA stages, the receiver

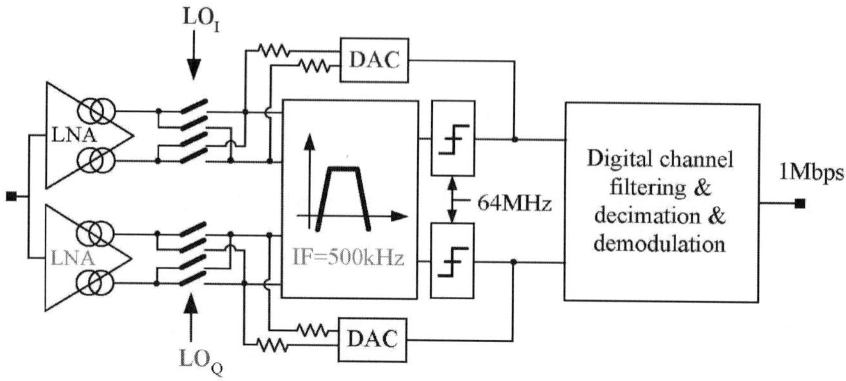

Figure 9.2: *Bluetooth receiver with digital demodulation, without preceeding analog filters or VGA*

hardly needs tuning and is much more flexible. On the other hand, though, this absence of analog conditioning has major implications on the ADC specifications:

- *high DR:* The ADC must handle the full signal range. The wanted signal can be very small because of the lack of preceding gain. Even then, the ADC must provide sufficient *SNR* to enable digital demodulation.
- *interferer immunity:* Interferers are present at the ADC input but they should not affect the resolution in the wanted channel. Therefore, the ADC must provide alias-suppression, it must be highly linear and interferers should not cause overload or spurious responses.
- *bandwidth requirements:* The interferer bandwidth can be 78MHz (corresponding to the bandwidth of operation for Bluetooth) while the sample rate of the ADC is only 64MHz. Hence, the bandwidth requirements are primarily set by the interferers.
- *termination of RF:* The ADC is directly connected to the RF front-end. It is in series with the passive mixer and acts as a load to the LNA. Since the LNA has a current-mode output the ADC must provide a low-impedance termination, i.e. its input impedance must be smaller than 400Ω over the entire 78MHz Bluetooth band.

9.2 Feed forward ΣΔ ADC

A feed forward ADC in combination with signal-conditioning in the decimation filter (see section 6.2) has been implemented for this Bluetooth receiver. As a short-hand notation, it is referred to as the "feed forward ADC". Further on, it is used as a reference to compare other conditioning ΣΔ ADCs to. At the output, a *SNR* of only 18dB is required for

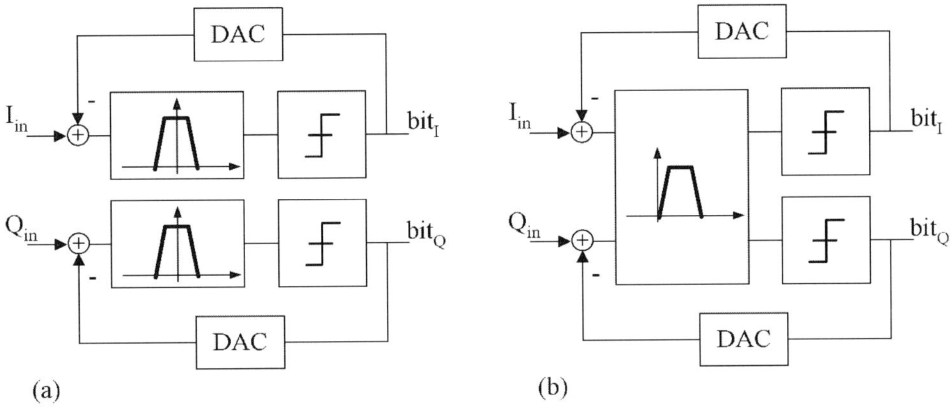

Figure 9.3: *Quadrature conversion using two low-pass ADCs (a) or using a single complex ADC (b)*

digital GFSK demodulation. Since the amplitude of the wanted signal may vary by 50dB and taking into account 5dB of margin, the ADC must then provide 73dB of *DR* in a bandwidth of 1MHz around the 500kHz low-IF frequency. The linearity requirement is set by the interferers. From the Bluetooth standard it can be derived that IM_3 distortion up to $-60dB_c$ can be allowed. Image rejection can be very moderate because the Bluetooth standard allows for an exception here: it is assumed that the image channel is not occupied by an interferer. *The major challenge on the Bluetooth ADC is in achieving a high DR at low power consumption while being robust to the presence of interferers.*

9.2.1 Design

Next to the above performance targets, system level choices in the receiver design set an additional boundary condition on the ADC architecture. For example, the system clock frequency is chosen as low as possible in order to avoid an additional PLL. Instead, it can be generated by an on-chip oscillator tuned at the third over-tone frequency of a crystal, generating 64MHz. Unfortunately, this clock frequency limits the over-sampling of the ADC. In order to meet the *DR* requirement either multi-bit quantization or very aggressive noise-shaping (i.e. a high-order loop filter) is needed. For reasons of linearity and power, the latter option is preferred (see discussion in section 3.2.1) and a 5th-order loop filter is used.

ADC architecture

The signal bandwidth from 0 to 1MHz can in principle be converted into the digital domain using two identical low-pass ΣΔ ADCs processing the I and Q channels separately (see fig. 9.3.a), just as for the FM/AM receiver in the previous chapter. The noise transfer

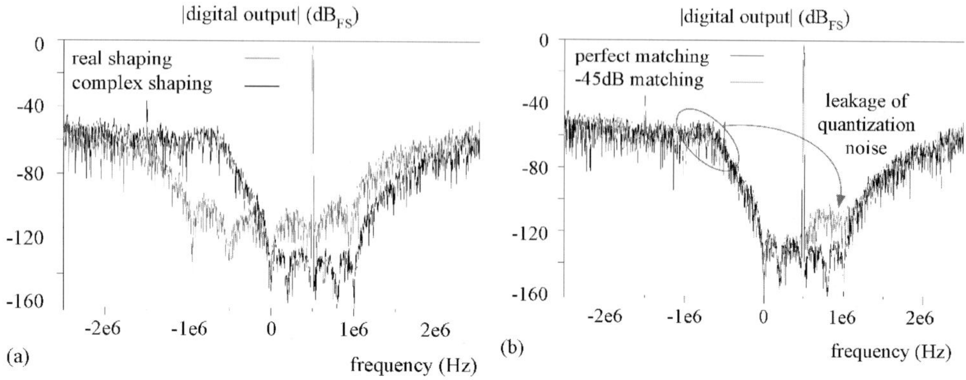

Figure 9.4: *Comparison between real (fig. 9.3.a) and complex noise-shaping (fig. 9.3.b) in graph (a) and effect of a small mismatch between the I and the Q path in graph (b)*

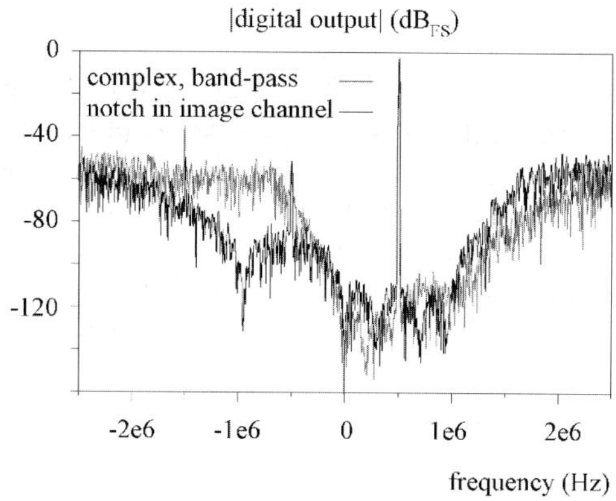

Figure 9.5: *Effect on noise-shaping when putting one notch in the image band*

function of the ADCs is symmetrical with respect to DC, in this case, and the same resolution is available from -1MHz to 0Hz as from 0Hz to 1MHz, which is not necessary. Hence, the idea to use a single complex ΣΔ ADC with quadrature inputs and outputs and with a poly-phase loop filter to achieve band-pass noise-shaping (see fig. 9.3.b). The noise-shaping of both topologies is compared in fig. 9.4.a. Complex noise-shaping is more efficient: because the quantization noise can be higher in the image band, it can be deeper suppressed in the wanted channel. In other words, the effective over-sampling has doubled. In view of the rather low sample frequency, complex noise-shaping seems particularly attractive.

One drawback of high-order, complex noise-shaping is in the risk of "leakage of quantization noise". The ideal, complex noise-shaping of fig. 9.4.b is only achieved in case of perfect matching between the STF of the I and the Q path of the ΣΔ ADC. In practice, mismatch causes quantization noise, present in the image channel, to leak into

Table 9.1: *Simulated SQNR of various 5th-order ΣΔ architectures (bandwidth is 1MHz, sample rate is 64MHz)*

	real (2 ADCs)	complex, matching is: perfect	-45dB	complex, 1 notch in image channel
SQNR	77dB	103dB	82dB	90dB

the wanted channel. Because of the very high noise at negative frequencies, this has a disastrous effect on the *SNR* in the wanted channel. Fig. 9.4.b illustrates this effect in case of a small gain (or phase) mismatch of only -45dB between the STF of the I and the Q path.

A solution to the leakage problem was introduced in [97]: instead of putting all filter notches in the wanted channel, one notch is left in the image channel (fig. 9.5). Obviously, the quantization noise in the wanted channel increases slightly but now, mismatch up to -20dB can be tolerated. Hence, the overall robustness of the architecture has improved considerably.

Table 9.1 compares the signal-to-quantization-noise ratio (*SQNR*) that is achieved by the various ΣΔ architectures. The complex noise shaping with one notch in the image band improves the *SQNR* by 13dB compared to the case when using two ADCs with real noise-shaping. This improvement is essential in order to achieve the 73dB target in the presence of other noise sources like circuit noise and clock jitter.

Fig. 9.6 shows the block diagram of the ADC with the 5th-order, complex loop filter. Starting from two low-pass filters with 90 degrees phase-shifted input signals, a band-pass filter is constructed by applying cross-coupling paths between the I and the Q filter at the output of the fourth and the fifth integrator. These cross-coupling paths realize complex filter coefficients and the filter response becomes asymmetric around DC. The notches appear at the following frequency:

$$\frac{g_{m,e4}}{2\pi C_4} \approx 750kHz \quad \text{and} \quad \frac{g_{m,e5}}{2\pi C_5} \approx 300kHz$$

The local feedback from the output of the third integrator to the second integrator introduces a pair of complex conjugated poles. This results in a notch at the edge of the signal channel and one at the edge of the image channel (i.e. at +1MHz and at -1MHz) and provides additional suppression of the quantization noise here. The resonance frequency can be calculated from eq. 8.7. Fig. 9.7 shows the simulated magnitude of the loop filter response. The finite output impedance of the various integrators limits the peak of the notches (i.e. reduces the quality factor of the resonance) and shifts the resonance frequency to a marginally lower value than what has been calculated.

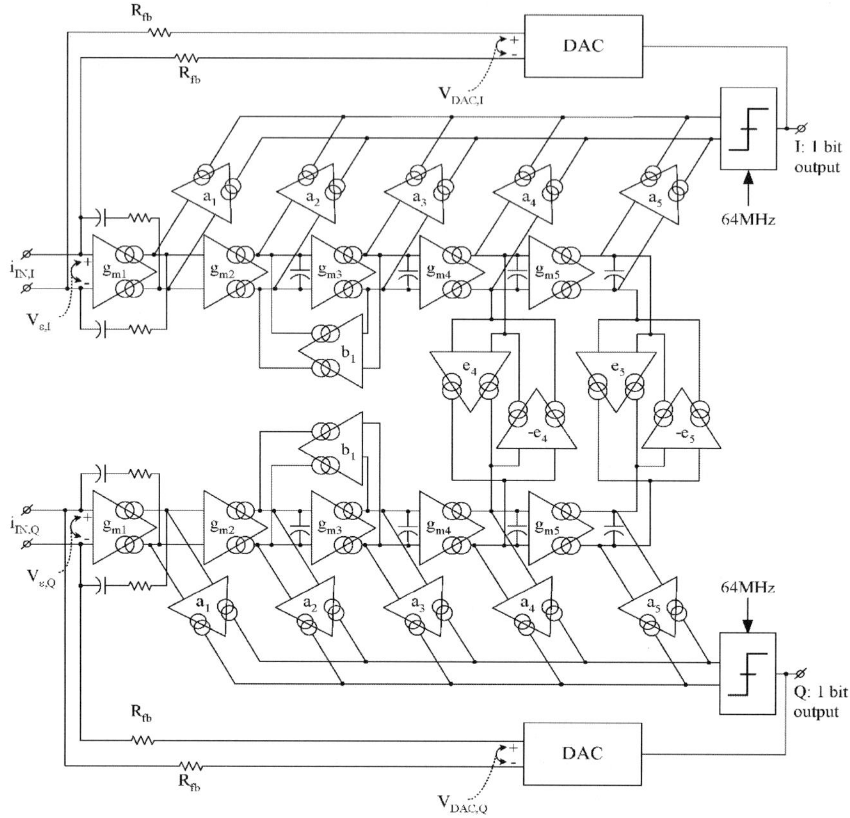

Figure 9.6: *5th-order ΣΔ ADC with complex loop filter*

ADC circuits

The design of the various OTAs is similar to the topologies of fig. 8.15 and fig. 8.16. These have been ported to a $0.18\mu m$-CMOS technology. Despite of the multiple stacked transistors this topology is successfully implemented within the 1.8V supply rail. Because it requires only a minimum number of current branches it remains preferred over a folded cascode OTA or any two-stage topology. Fig. 9.8 shows the nominal bias voltage of various nodes and demonstrates the available headroom.

Contrary to the implementation in section 8.3 in the $0.25\mu m$-technology, here, N-MOS devices are used for the input transistors. In the $0.25\mu m$-technology, with a low-impedance substrate, P-MOS devices were chosen for isolation reasons. The $0.18\mu m$-technology uses a high-impedance substrate (i.e. the resistivity equals $10\Omega cm$) and therefore this argument becomes less important. On the other hand, at the same quiescent current, the N-MOS transistor yields a three times higher g_m than the P-MOS transistor

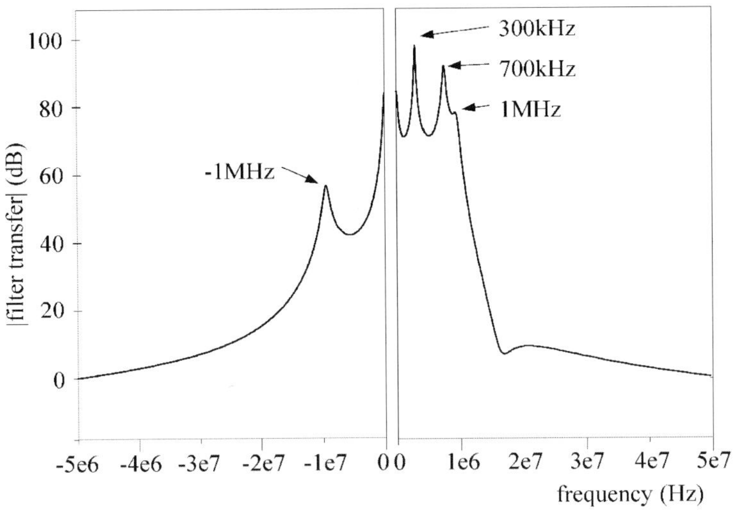

Figure 9.7: *Magnitude of loop filter transfer*

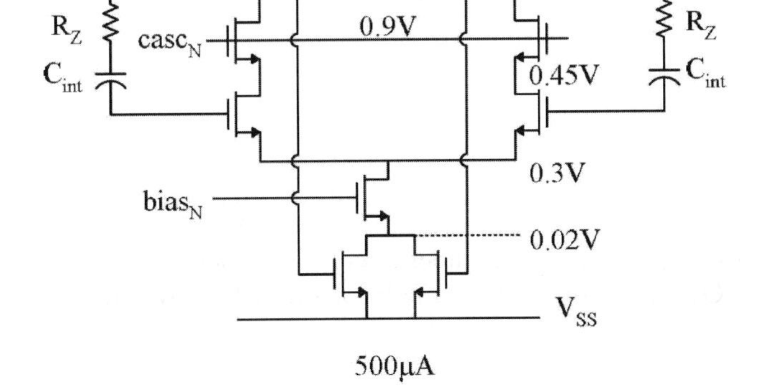

Figure 9.8: *DC biasing of first integrator*

Table 9.2: *Variation on component values in the* $0.18\mu m$*-CMOS technology*

Processing spread		Temperature [5; 125] deg C	
N^+-poly resistance	gate ox. capacitance	N^+-poly resistance	gate ox. capacitance
$4\sigma = 17\%$	$4\sigma = 10\%$	$[-1\%; +3\%]$	$< 1\%$

and is preferred for this reason. Since the N-MOS transistor is referenced to ground (via its bulk connection), all other signal-conducting devices are referenced to ground as well such that any substrate bounce remains common mode.

The time-constants of the loop filter depend on the product of a N^+-poly resistance and gate-oxide capacitance. Table 9.2 lists the expected variation on these parameters. The overall, worst-case spread on the RC-product is less than 25%. In the FM/AM ADC, the spread was dealt with by providing a bandwidth margin. Here, the spread is compensated for by adding or removing integrator capacitors in discrete steps. Therefore, the filter bandwidth is made programmable in four steps allowing a correction of -12.5% up to $+25\%$. Note that this interval is asymmetric. Especially the case where the filter bandwidth is smaller than intended needs correction since, then, the quantization noise shifts into the wanted channel. The calibration algorithm is out of the scope of this book.

Fig. 9.2 implies that the ADC has a current-mode input. The input is directly driven by the output current of the LNA that is down-converted by the passive mixer (similar to the mixer in the FM/AM receiver and discussed on page 126). The ADC must provide a low-impedance termination of the RF front-end to guarantee a linear operation of the LNA and the mixer. It is calculated that the differential input impedance of the ADC must be below 400Ω over the entire bandwidth of the input signal, including interferers (i.e. over 78MHz). The differential input impedance of the ADC is calculated:

$$Z_{in}(j\omega) \approx \frac{2}{\frac{F(j\omega)}{R_{fb}} + g_{m1} + j\omega C_{gate}} \tag{9.1}$$

Most parameters in this formula have been defined in fig. 9.6, others are:

$F(j\omega)$ = transfer function from $v_{\epsilon,Q}$ to $v_{DAC,Q}$ in fig. 9.6 (or from $v_{\epsilon,I}$ to $v_{DAC,I}$)

C_{gate} = gate capacitance of first OTA

At low frequencies, the input impedance approximates $2R_{fb}/F(j\omega)$ and is very small due to the high loop gain. Beyond the unity-gain bandwidth of the loop the input impedance

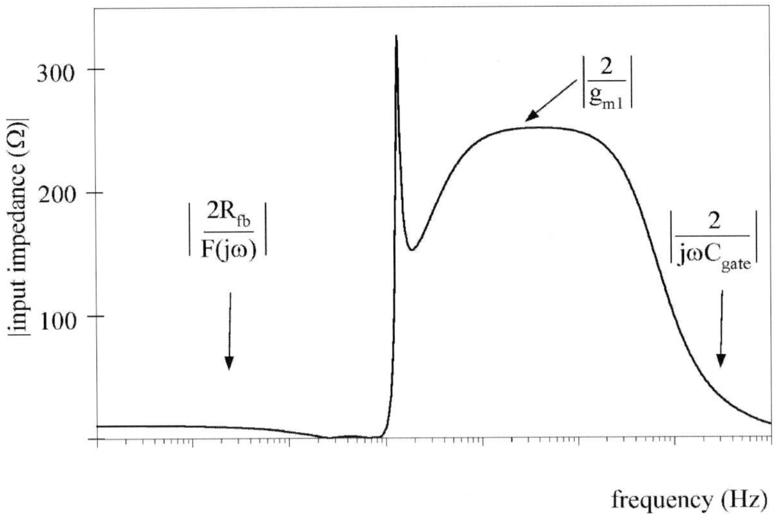

Figure 9.9: *Differential input impedance of ADC as a function of frequency (peaking is due to numerical inaccuracy)*

of the ADC equals that of the first stage, i.e. $2/g_{m1}$. For very high frequencies, Z_{in} is limited by the gate capacitance of this stage. The magnitude of Z_{in} is plotted in fig. 9.9 for positive frequencies. The differential input impedance remains below 400Ω for all frequencies.

In this particular receiver architecture, the requirement for a low-impedance termination sets the quiescent current of the first stage and dominates the power consumption of the overall receiver. This quiescent current is so high that -contrary to the generic assumption in the power/performance relation of eq. 4.13- here, the noise and distortion of the first stage is less important than that of the feedback resistor R_{fb} and of g_{m2}. This would not be the case for an optimized, stand-alone ADC.

The circuit noise is significantly larger than the quantization noise and limits the achievable *DR* to 79dB. On top of that, jitter on the DAC clock translates into an additional noise contribution. The short-term jitter of the 64MHz-oscillator amounts $7ps_{rms}$ (from measurements). Assuming white noise and based on [7] it can be calculated that, due to the jitter, the overall *DR* of the ADC further degrades to 76dB.

Digital conditioning and demodulation

Fig. 9.10 shows a functional diagram of the digital conditioning blocks and the demodulation. A CIC filter (see eq. 4.17 with $k=6$) performs decimation filtering and down-sampling by a factor 8. The output sample rate equals 8MHz. A rotating CORDIC-block [106] shifts the IF frequency of the outputs of the CIC filters from 500kHz to DC. The consequent matched filter provides optimal suppression of white noise. In addition, it also

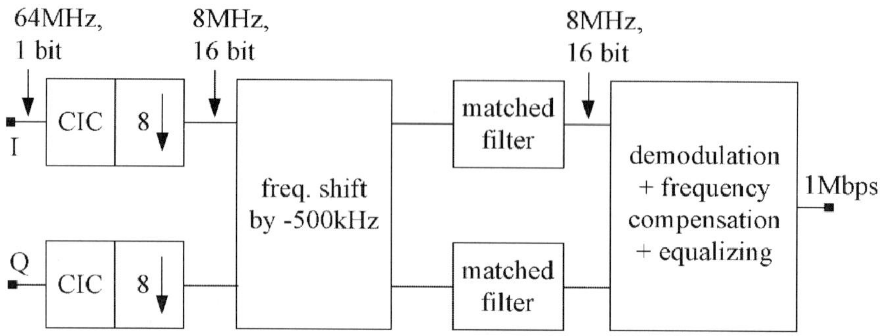

Figure 9.10: *Functional diagram of digital part of the receiver of fig. 9.2*

suppresses the remaining quantization noise from the ΣΔ ADC and therefore can be considered as a decimation filter as well. The remaining digital blocks perform demodulation and compensation of a potential frequency offset [121]. These functions are beyond the scope of this book.

9.2.2 Evaluation

The ADC has been embedded in a single-chip Bluetooth receiver (see fig. 9.2), designed in a $0.18\mu m$ digital CMOS technology. Fig. 9.11 shows the die micrograph with the various receiver blocks. The evaluation of the overall receiver is documented in [121]. Highlights are the sensitivity of -71dBm at a power consumption of less than 32mW, the area of $3.5mm^2$ and the IIP_3 of -11dBm achieved without any analog filtering or VGA.

Here, the focus is on the performance of the ADC, though. The evaluation of a stand-alone version is presented. The test IC includes the complex ΣΔ ADC, a bandgap reference, the crystal oscillator and low-swing buffers to output the bitstreams. Resistors of $10k\Omega$ are added in series with the input terminals to facilitate evaluation. This value is comparable to the output impedance of the RF front-end. While in the actual receiver the maximum differential input signal to the ADC (i.e. the output of the mixer) is $100\mu A_{pp}$, it equals $1V_{pp}$ in this set-up.

Set-up

An Arbitrary Waveform Generator (Tektronix AWG420 with 16-b word-length and a sample rate up to 200MSps) is used to generate the differential, quadrature input signals with sufficient resolution. For some tests, external calibration of the generator is required to reduce harmonic and intermodulation distortion of the generator. For tests with high-frequency interferers, high-pass filters (Mini-Circuits BHP-25) with stop-band until 25MHz are used to suppress the low-frequency spurious from the generator. The bitstream outputs are grabbed with a VXI-analyzer and further processed in LabView.

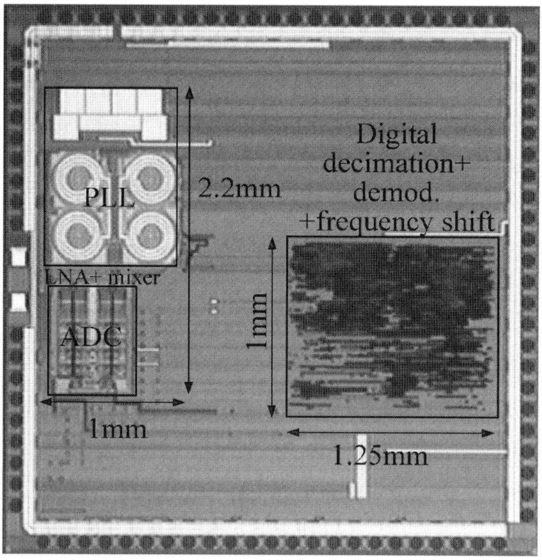

Figure 9.11: *Micrograph of the single-chip Bluetooth receiver*

Conventional performance parameters

First, some conventional ADC performance parameters such as *SNR*, linearity and image rejection are discussed.

Conventional performance parameters: Fig. 9.12.a shows the *SNR* as a function of the digital output: a peak-*SNR* of 75.5dB and a *DR* of 76dB are achieved. Fig. 9.12.b shows the result of a two-tone test: IM_3-distortion is below $-82dB_c$ relative to the input tones, each at half of the full-scale input. IR is typically better than 50dB and is mostly dominated by a gain mismatch. The worst-case value, measured over 17 samples is 47dB. This is far better than required. Further statistics on the measured image rejection are quoted in [32].

The target specifications as defined in the introduction, are all well met. Table 9.3 gives an overview of the conventional ADC performance parameters and evaluates the FOM_{SINAD} of eq. 4.4. The latter is discussed in section 9.5.

Interferer immunity

Because of the lack of preceding filters in this particular receiver topology, the ADC immunity with respect to interferers must be evaluated next.

Aliasing limit on interferer immunity: In a first test, an interferer is applied near the sample frequency of the ADC in order to evaluate the aliasing suppression. The input level of the interferer complies with the blocker level defined in the Bluetooth standard [112].

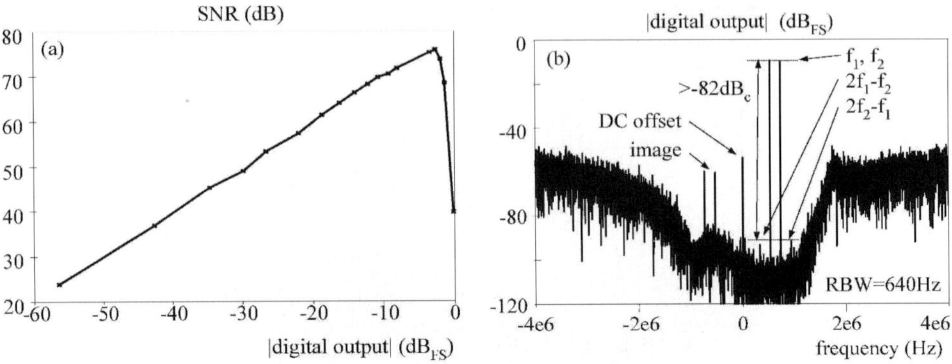

Figure 9.12: *SNR as a function of digital output (a) IM_3-test with f_1=530kHz and f_2=730kHz (b)*

Table 9.3: *Measured performance of the complex ADC (conventional parameters)*

full-scale input	$100\mu A_{pp}$ (differential)
bandwidth	0-1MHz
sample rate	64MHz
DR	76dB
peak-*SNR*	75.5dB
IM_3	$<-82dB_c$
IR (17 samples)	$>47dB$
$I_{quiescent}$	2.5mA @1.8V
active area	$0.22mm^2$
technology	$0.18\mu m$-CMOS, 1P, 5Al
FOM_{SINAD} eq. 4.4	$1 \times 10^{-16} J$

The interferer frequency is chosen slightly higher than the clock frequency such that the alias appears at a positive frequency in the bandwidth of the wanted signal (fig. 9.13). (In case the interferer frequency was lower than the sample frequency, the alias would appear at a negative frequency.) The alias component is below the allowed limit of $-68dB_{FS}$ derived from the Bluetooth blocker tests[1]. In general, for all measured samples, the aliasing

[1]The test defines a blocker at -27dBm, corresponding to -10dB of the ADC full-scale output, while the wanted signal is at -67dBm, corresponding to -50dB of the ADC full-scale output. Since 18dB of *SINAD* is

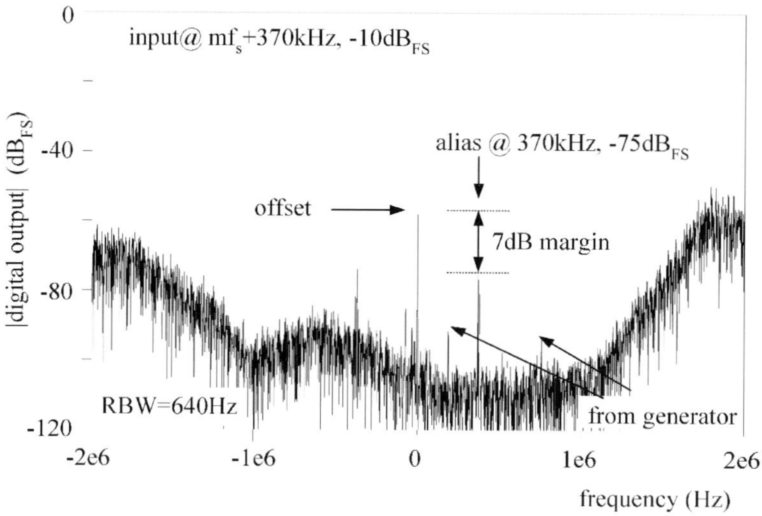

Figure 9.13: *Aliasing test with an input signal at +370kHz offset from the sample frequency at a signal level of* $-10dB_{FS}$

suppression exceeds 60dB. Still, this is lower than what is expected from eq. 6.3. This discrepancy is probably due to the fact that -next to the aliasing- also parasitic mixing via cross-coupling occurs. As such, the interferer is mixed back into the signal band.

Stable input range for interferers: A second test concerns the stable input range for interferers. As discussed in section 6.1.3, this limit is frequency dependent, being inversely proportional to the STF of the ADC. The simulated limit as well as the measured stable input range over frequency is depicted in fig. 9.14. The simulated limit is based on an AC analysis of the implemented ADC where DAC and quantizer have been modeled by a linearized gain. The simulated limit deviates from the measured curve, especially for adjacent channels. This is due to numerical problems caused by a fast transition of the amplitude and the phase of the loop gain. In addition, the simulated limit is based on a linearized model assuming small-signal operation. This model becomes inaccurate near full-scale and, therefore, should be considered as a first-order estimate only. Better correspondence can be achieved if transient simulations are used to predict the stable input range. The penalty of that approach is in a very long simulation time.

The measured graph is asymmetrical around DC. This is due to the complex nature of the loop gain and the STF. At positive frequencies, the stable input range is large enough to accommodate the interferers defined in the various Bluetooth tests and indicted as the “Bluetooth mask”. At negative frequencies, the margin is tight, though. For this test, only five samples have been evaluated because the frequency sweep needs to be performed manually. (Because of the differential quadrature inputs, an automatic set-up would re-

required for demodulation, the alias needs to be below $-68dB_{FS}$.

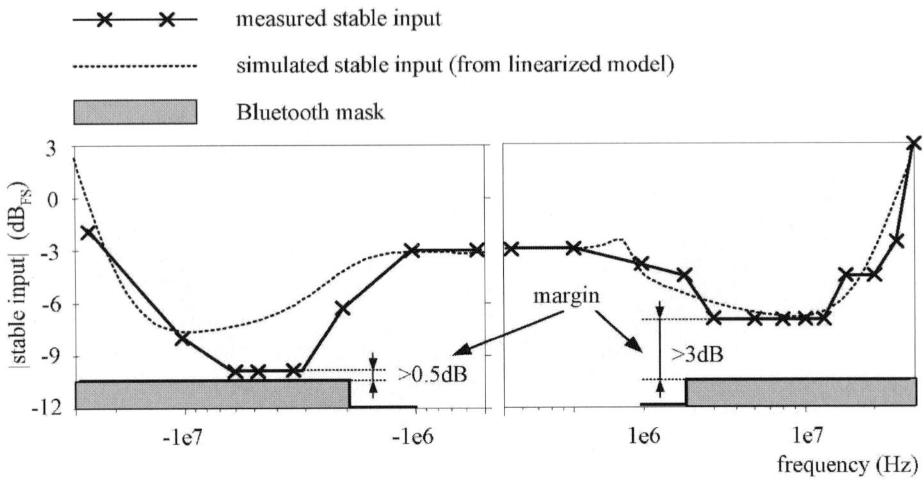

Figure 9.14: *Stable input range over frequency*

quire four programmable attenuators of sufficient performance. These were not available at the time.) The variation on the measured stable input range is small. All measured samples accommodate the Bluetooth mask as depicted in fig. 9.14.

Spurious responses due to interferers: Thirdly, an interferer around half the sample frequency, i.e. 32MHz is applied. In terms of spurious responses, this is a worst-case input since the ΣΔ modulator tends to make correlated patterns at this frequency. It is evaluated that, as long as this interferer remains smaller than 0.1V (this corresponds to -17dB on the digital scale), the spurious responses within the wanted channel do not affect the *DR* of 76dB.

IM_3 distortion of interferers: Finally, IM_3 distortion is evaluated over frequency: interferers are applied at frequency f_1 and f_2, such that the intermodulation component $2f_1-f_2$ falls in the middle of the wanted channel, i.e around 500kHz. The interferers f_1 and f_2 are applied at an input level equaling half the maximum stable input at those frequencies. This input level may differ from that of a conventional IM_3-test for wanted signals where two inputs at $-9dB_{FS}$ (adding up to $-3dB_{FS}$) are applied. In case interferers were applied at the latter input level, they would either cause compression (in case the stable input range is lower than $-3dB_{FS}$) or the IM_3-components would be hidden in the noise (in case the stable input range corresponding to this input frequency is much higher than $-3dB_{FS}$).

It was mentioned in section 4.4 that the transconductors of the input stage are the dominant source of distortion in a feed forward ADC. This is true for the wanted channels. Fig. 6.16 shows that, especially when applying adjacent interferers, the internal signal swings can become large. For these frequencies, distortion at the output of the first integrator may become important as well. This is evaluated next. Afterwards, IM_3-

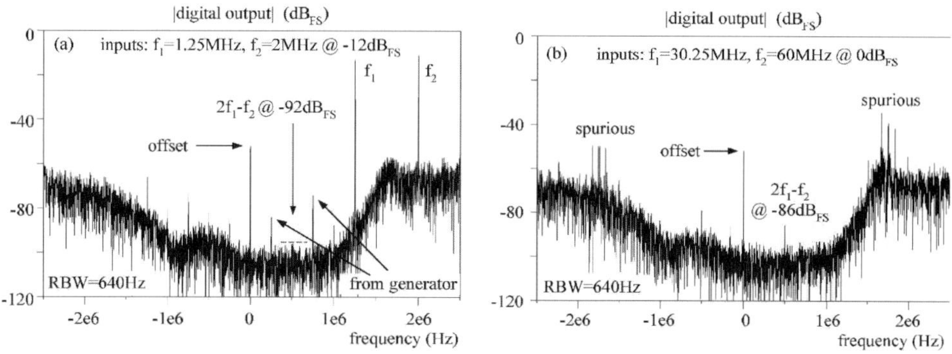

Figure 9.15: *IM_3 test with adjacent interferers: f_1=2MHz, f_2=1.25MHz (a) and with far-off interferers: f_1=30.25MHz, f_2=60MHz (b)*

distortion of far-off interferers is measured.

Fig. 9.15.a shows the IM_3-distortion when applying *adjacent interferers* f_1=*1.25MHz* and f_2=*2MHz*, each at $-12dB_{FS}$. As such, the interferers add up to the maximum stable input of $-6dB_{FS}$ for this frequency range. The distortion component at 500kHz is below $-92dB_{FS}$. Note the low-frequency spurious components at 250kHz, 750kHz, etc. These are due to the generator. The component generated at 500kHz has been suppressed by calibration of the generator such that only the contribution of the ADC appears in the output.

As mentioned, distortion is lower for *far-off interferers* [112]. This is shown in fig. 9.15.b with inputs at f_1=*30.25MHz* and f_2=*60MHz*. In this set-up, high-pass filters with the stop-band up to 25MHz suppress the low-frequency spurious generated by the AWG. Again, the IM_3 component falls at 500kHz. Both inputs are applied at $0dB_{FS}$ and add up to $+6dB_{FS}$. (This is 9dB higher than in a conventional measurement because, on one side, the stable input range is high enough to allow for this input level and, on the other side, applying a signal at the conventional input level would require too many samples to detect the IM_3 component). Clearly, the distortion is low enough not to limit the allowable interferer level for these frequencies. In fact, the allowable input for far-off interferers is limited by the supply voltage. Also note the tones around 1.75MHz (and -1.75MHz): these spurious components are caused by the large input near half the sample frequency; i.e. these are spurious responses. They must be suppressed in the decimation filter.

Fig. 9.16 summarizes the evaluation of the interferer immunity of the presented reference feed forward ADC. For reference purposes, it is repeated that the input amplitude of a full-scale *wanted* signal equals $0.5V$, corresponding to $-3dB$ on the digital scale. This graph corresponds to the simulations on the 3rd-order ADC discussed in chapter 6 (fig. 6.8). *For adjacent channels, the allowable interferer level is limited by the stable input range of the ADC. Around 32MHz, the interferer level must not exceed 0.1V (this corresponds to -17dB on the digital scale) to maintain a DR of 76dB in the wanted chan-*

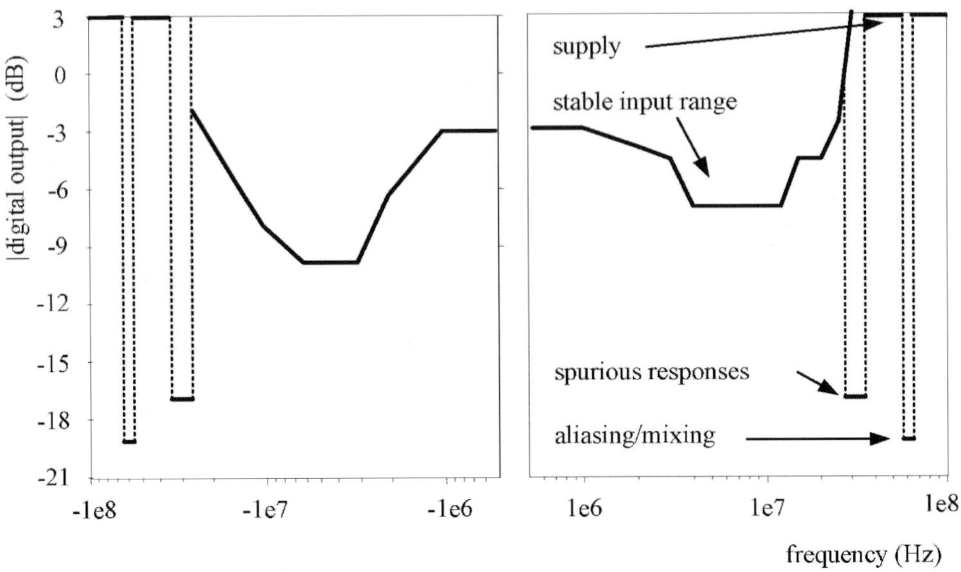

Figure 9.16: *Overall limit on the allowable interferer level over frequency*

nel. Around 64MHz, interferers up to $-19dB_{FS}$ are allowed: in view of the aliasing or parasitic mixing of $-60dB$ this input level does not affect the DR of 76dB. For all other frequencies, the allowable interferer level is set by the supply voltage. This limit corresponds to $+3dB$ on the digital scale; i.e. an interferer with an amplitude of $1V$. Note, distortion is not a limiting factor for the interferers applied to this ADC.

9.3 Conditioning feedback ΣΔ ADC

In this section, a ΣΔ ADC with feedback compensation is presented as an alternative solution for the receiver baseband of fig. 9.2. The feedback topology is used in view of its filtering STF and allows signal-conditioning integrated into the ADC (see section 6.3). The feedback ADC targets the same performance as the feed forward ADC in the previous section and has been implemented in the same $0.18\mu m$ CMOS technology. Moreover, the same circuit topologies have been reused for the various building blocks.

The asset of the feedback ADC is in its improved interferer immunity. The presented design illustrates the theory discussed in section 6.3. In particular, it clarifies the trade-off between:

- an improved interferer immunity;
- a competitive power/performance ratio.

This trade-off is reflected in the discussion in section 9.3.1 and is is evaluated based on measurement results in section 9.3.2.

9.3.1 Design

Since the feedback ADC targets the same *DR* as the feed forward reference ADC, essentially, the same NTF needs to be realized. In view of interferer immunity, though, a different -i.e. filtering- STF is pursued. It is shown that, in order to come to a power-efficient implementation of the feedback ADC, some concessions to the ideal NTF and STF must be made. Afterwards, some circuit details are presented.

ADC architecture

The feedback ADC can be designed starting from the same NTF as for the feed forward reference ADC of section 9.2, i.e. the same function L_1 (see chapter 6) is realized but it is mapped on a filter with feedback compensation. In section 6.3, it was shown that this yields a filtering STF but it also results in unfavorable values for the unity-gain frequencies of the integrators in the loop filter.

Example

Assuming the same boundary conditions as for the feed forward design (i.e. the same NTF, the same maximum input signal and the same allowable swing at the integrators' output) a unity-gain frequency of 300kHz is required for the first integrator in the feedback topology compared to 2.8MHz in the feed forward topology. This implies that in the feedback topology the first integrator has attenuation inside of the signal bandwidth of 1MHz. Obviously, this is very disadvantageous for the overall power/performance balance.

This problem is reduced by making the concessions listed next.

- *A* 4th *order instead of a* 5th *order loop filter is implemented*: the unity-gain frequency of the first integrator can then increase to 600kHz. (Since less feedback paths are present and thus less signal is injected in the loop the first integrator can have higher gain.) The penalty of this choice is in a reduced *SQNR* of about 80dB instead of 90dB for the 5th order design. In addition, the interferer immunity is slightly affected since most of the limitations discussed in section 6.1.3 depend on the loop filter order.

- *A local feed forward path replaces a feedback path*: this is illustrated for a simple 3rd order example in fig. 9.17.a and b. It can be calculated that both topologies are equivalent if the local feed forward coefficient equals d_2/d_1 and if the first and the second integrator are interchanged. As such, the unity-gain frequency of the first integrator becomes 1.9MHz. The penalty of this adaptation is in a further reduction of the slope of the filtering STF by one order and in some overshoot in the adjacent channel. In addition, the feed forward path reduces the aliasing suppression by one order and affects the immunity to spurious responses.

Both measures compromise on the interferer immunity but improve the power/performance balance by allowing more gain in the first integrator of the loop filter. (Note that the gain of the first integrator is still lower than in the feed forward implementation.) This

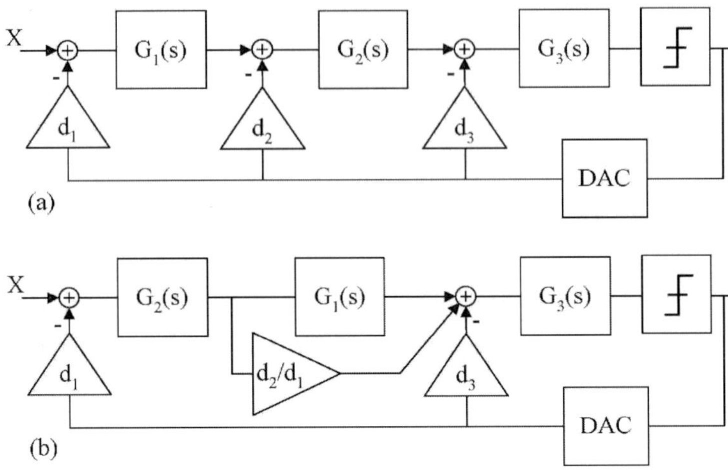

Figure 9.17: *Conventional feedback topology (a) and modification with a local feed forward path (b)*

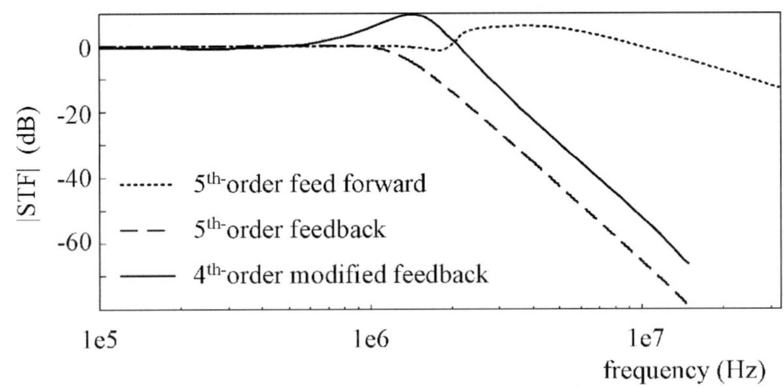

Figure 9.18: *Simulated STF of a 5th-order feed forward, 5th-order feedback and a 4th-order modified feedback ΣΔ ADC*

compromise is evaluated in section 9.3.2 when presenting the measured performance. For now, fig. 9.18 compares the STF of the presented topology to that of the 5th-order feed forward ADC of the previous section and that of a 5th-order feedback ADC. These are simulated curves based on a linearized model of the various topologies. The filtering behavior of the present topology is not as good as that of a 5th-order implementation with "conventional feedback". It is still an improvement on that of the feed forward topology.

Fig. 9.19 depicts a block diagram of the implemented, "modified" feedback ADC. As in the previous ADCs, some local resonators have been implemented to create additional loop gain at particular frequencies. The pair of complex conjugated poles hence creates a notch at +1MHz and at -1MHz (for suppression of quantization noise in the image

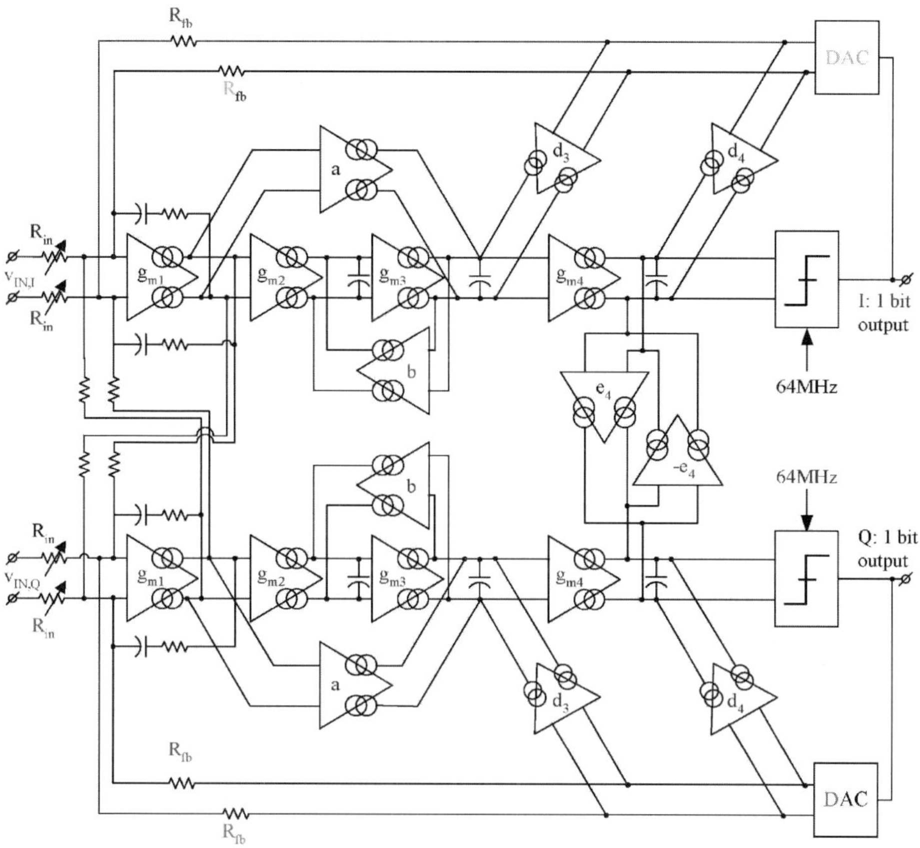

Figure 9.19: *4th-order ΣΔ ADC with modified feedback topology*

channel). The cross-couplings between the I and the Q path of the loop filter create notches at 200kHz and at 600kHz. The variability of R_{in} is discussed next. The variability of R_{in} is discussed next.

Integration of programmable gain

Because of the order of its loop filter, this ADC achieves a lower *SQNR* than the feed forward ADC in the previous section. In addition, the thermal noise of this ADC is somewhat higher in order to save on power consumption. By consequence, *the design of fig 9.19 has a lower DR, and a lower peak-SNR. However, this topology excels in interferer immunity. Hence, the option to extend its DR by integrating some programmable gain:* the input resistance is switched from $100k\Omega$, to $10k\Omega$ and to $1k\Omega$ depending on the amplitude of the wanted signal[2]. This results in a dynamic adaptation of the input range to the strength of the wanted signal and in a larger input-referred *DR*.

[2]Information on the level of the wanted signal is available in the digital domain, after decimation. Alternatively, it can be derived from the modulation depth of the bitstream signal at the output of the ADC [122].

Table 9.4: *Simulated SNR and contribution of quantization and of thermal noise*

R_{in}	$v_{in,max}$	$SQNR$	SNR_{th}	overall SNR
$100k\Omega$	$0.35V_{rms}$	80dB	71.5dB	71dB
$10k\Omega$	$0.035V_{rms}$	80dB	65dB	65dB
$1k\Omega$	$0.0035V_{rms}$	80dB	50dB	50dB

Note that only the wanted signal is used for the gain control. Event hough interferers are present and even while they may be much stronger than the wanted signal, the interferers do not cause overload because the ADC has attenuation at these frequencies. This assumes that the limit on the allowable interferer (see evaluation in the next section) level is not exceeded. A similar assumption would be required for a conventional architecture based on a cascade of filters, programmable gain sections and an ADC: the programmable gain control can be based on the strength of only the wanted signal if the interferers have been suppressed sufficiently in the preceding filter.

The programming of the input resistance does not interfere with the noise shaping and therefore the *SQNR* of the ADC is the same in all three settings. The overall *SNR* including circuit noise, however, does depend on the value of R_{in}. This is due to the following reasons:

- the thermal noise of R_{in} changes per setting;
- the input-referred thermal noise contribution of the other blocks changes because of a different transfer of these sources;
- the signal power to which the noise is referred to, changes per setting ($v_{IN}=0.35V_{rms}$ for $R_{in}=100k\Omega$ while $v_{IN}=3.5mV_{rms}$ for $R_{in}=1k\Omega$).

Table 9.4 lists a summary of the expected *SNR* due to quantization noise only (*SQNR*), due to thermal noise only (SNR_{th}) and the overall *SNR* including both contributions.

ADC circuits

In order to avoid overlap with previous discussions, this section does not aim at a systematic and complete analysis of all circuits. Instead, a few distinct items, specific to the feedback implementation are highlighted.

Noise budget: Despite of the efforts to increase the unity-gain frequency of the first integrator, its gain at the edge of the signal bandwidth is still moderate (i.e. the unity-gain

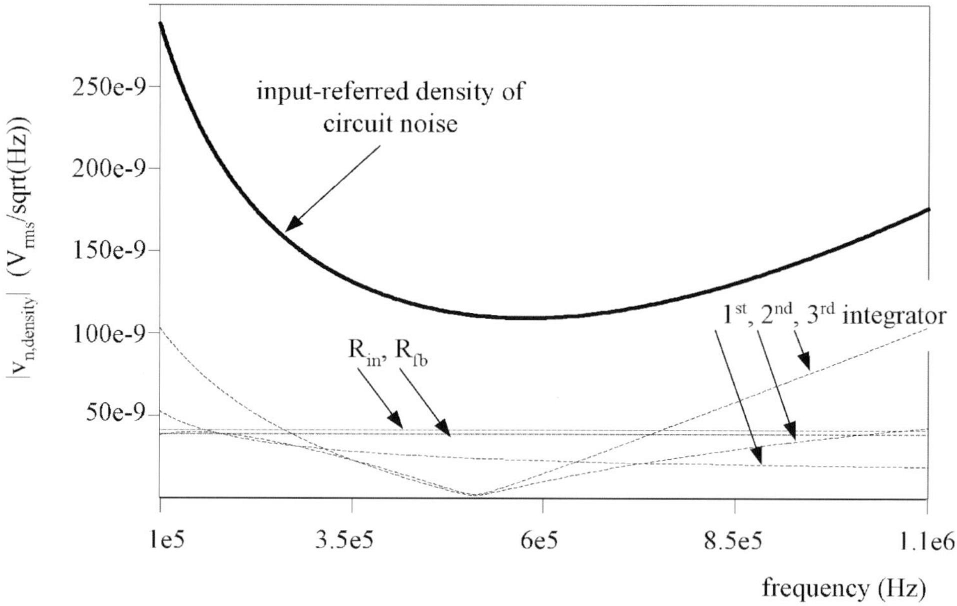

Figure 9.20: *Input-referred noise density of the implemented feedback ADC with dominant contributions*

frequency of the first integrator equals 1.9MHz while the signal bandwidth is 1MHz). As a consequence, the second and the third stage of the loop filter contribute to the overall noise of the ADC (see fig. 9.20). This was not the case in the feed forward design. Hence, here, all integrators (except for the first one consuming $500\mu A$) are biased at $200\mu A$ while in the feed forward ADC they were biased at $100\mu A$. For use in section 6.3.3, it is derived that the implementation factor of the feedback design, as compared to an equivalent feed forward design equals: $F = (500 + 4 \cdot 200)/(500 + 4 \cdot 100) \cong 1.5$

Distortion at internal nodes: In the feed forward reference ADC, the transconductors of the input stage represent the dominant source of distortion. In a feedback implementation, the consequent stages (i.e. the second stage, the third stage, etc.) also contribute. This is due to the following reasons:

- the effect of distortion at these nodes on the overall distortion is larger than in the feed forward implementation because the preceding loop gain is lower;
- the amount of distortion at the internal nodes is larger than in the feed forward case because of a larger signal swing.

The latter has been illustrated in figures 6.15 and 6.16 in section 6.3 for a 3rd-order conventional feedback ADC. However, the local feed forward path in the modified feedback implementation changes the transfer function from the input of the ADC to the output of

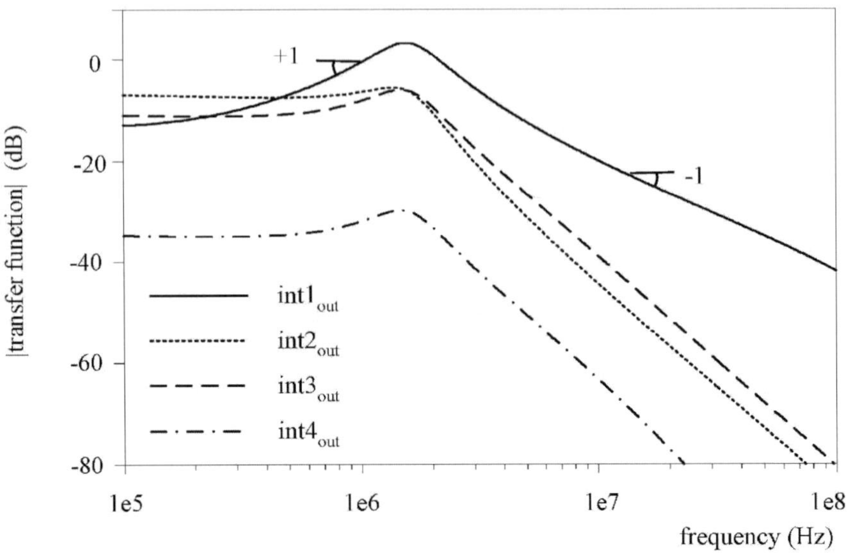

Figure 9.21: *Transfer function from the input of the ADC to the internal nodes based on a linearized model of the modified feedback ADC*

the first integrator. This is shown in fig. 9.21. In view of distortion, the signal swing at the output of the first stage is most important. The frequency dependence of this signal swing results in a strong frequency dependence of the IM_3-distortion of this ADC as well. For example, in the wanted channel (i.e. from DC to 1MHz) the following frequency dependence is expected:

- at the output of the first stage, the IM_3-distortion as a function of frequency has slope +2 because the fundamental signal has slope +1 (see eq. B.2);
- the input-referred IM_3 shows a slope of +3 because the transfer function from the output of the first stage to the input of the ADC has slope +1 (i.e. the inverse of the first integrator).

The frequency dependence of the input-referred IM_3 for adjacent or far-off interferers can be anticipated in the same way.

Graceful degradation: In case a large input signal is applied to a ΣΔ ADC, the internal signals grow and the integrators may saturate. In case of a feed forward implementation, the forward paths by-pass the saturated stages and keep the loop functional. This is referred to as "graceful degradation". Even though the *SQNR* drops dramatically, the loop remains operational and is able to recover fast when the input signal becomes smaller again.

In a feedback implementation, this does not happen: the saturated integrators block the loop operation because they are not by-passed. For the same reason, it takes a long

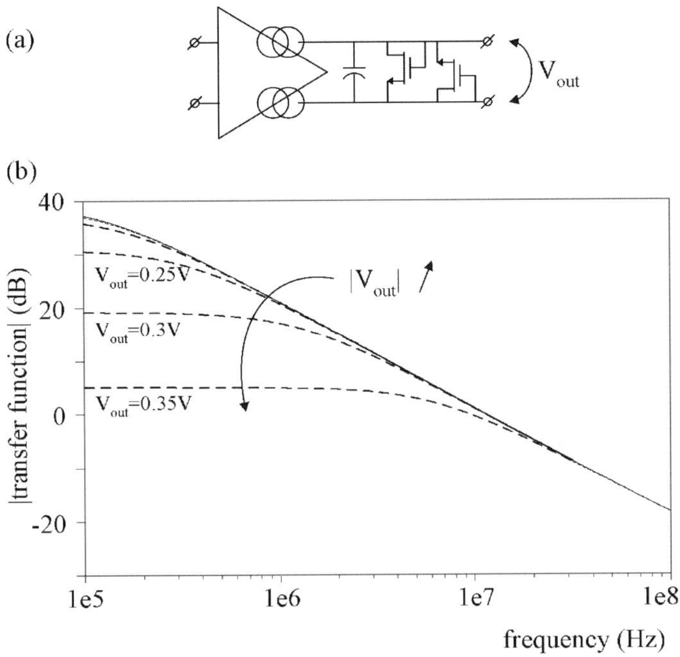

Figure 9.22: *Implementation of "graceful degradation" (a) and transfer function with output voltage as a variable (b)*

time to recover from an over-load condition. Therefore, an alternative implementation of the "graceful degradation" technique is required. Fig. 9.22.a shows the implemented solution: two diodes (M1 and M2) with reverse polarization are connected between the outputs of the integrator. The threshold voltage of the diode transistors is about 250mV. If the output swing exceeds this value, the diodes become a low-impedance load for the g_mC-integrator. Instead of saturating, the g_mC-stage starts acting as a low-gain amplifier (see fig. 9.22.b). Therefore, the order of the loop filter and of the noise-shaping reduces while the modulator remains stable. As soon as the signal swing becomes smaller again, the diodes go in the "off"-condition and the integration is resumed.

9.3.2 Evaluation

This ADC has been embedded in a test IC similar to the previous one; i.e. it has on-chip clock generation, a bandgap reference, etc. The set-up for the evaluation is the same as discussed before. First, the conventional performance parameters are evaluated. Next, the interferer immunity is discussed.

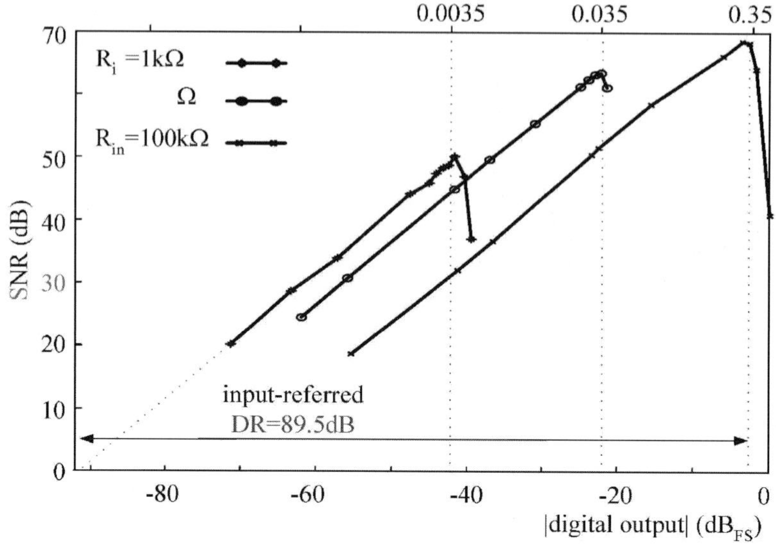

Figure 9.23: *SNR versus input swing for the three settings of R_{in}*

Conventional performance parameters

Fig. 9.23 shows the *SNR* as a function of the digital output for the three settings of the input resistance. Because of the programmable gain, the input-referred *DR* has extended to 89.5dB. The peak-*SNR* remains moderate because of the simplified loop filter topology. This results in relaxed specifications on the consequent blocks in the conditioning channel and on a number of reference circuits (see section 9.5). Note that the peak-*SNR* and the *DR* range differ per setting. This is due to the fact that the relative contribution of thermal noise -compared to quantization noise- differs for the various R_{in} settings (see table 9.4).

This is also shown in the output spectra of fig. 9.24.a and b: for $R_{in}=100k\Omega$ the shaping of the quantization noise is clearly visible. For $R_{in}=1k\Omega$ it is covered by an almost white noise floor due to the thermal noise. This white noise floor even hides the "bump" of quantization noise that is present in the image channel in fig. 9.24.a.

Fig. 9.25 shows the IM_3 distortion in the wanted channels as a function of frequency. Two nearby signals, i.e. at frequency f and f+30kHz are applied and the frequency f is swept. As explained above, in the feedback ΣΔ ADC, dominant distortion occurs especially at the output of the first stage and, within the wanted channel, has a slope of $18dB/oct$ as a function of frequency (see discussion on page 161). Finally, note that the measured IM_3 is very similar for all settings. Even though the input voltage differs in the three settings, the input current is always the same because R_{in} is scaled. By consequence, also the internal signal swings -determining the distortion- are the same.

Table 9.5 gives an overview of the conventional ADC performance parameters. Evaluating the FOM_{SINAD} of eq. 4.4 seems to indicate that the ADC does not perform well,

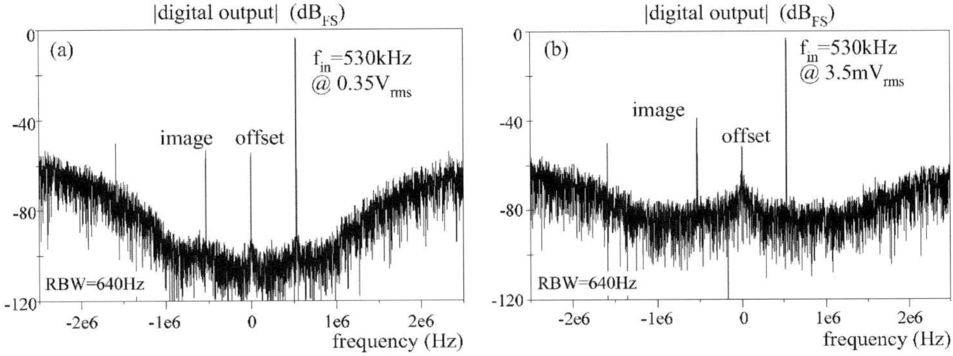

Figure 9.24: *Full -scale input for $R_{in} = 100\Omega$ and for $R_{in} = 1k\Omega$*

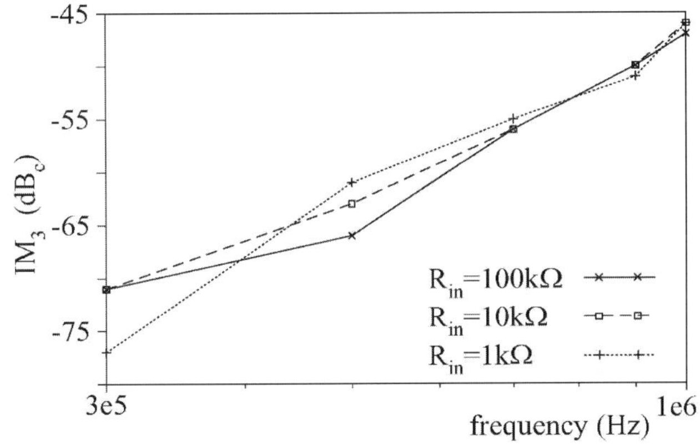

Figure 9.25: *IM_3 of wanted channels over frequency*

in any of the settings, as compared to the ADCs listed in the appendix A. This demonstrates a shortcoming of the use of FOMs: they are only suited for the assessment of a stand-alone ADC. Here, the ADC includes signal-conditioning; i.e. it handles a very large input-referred *DR* and, at the output, only provides a moderate *SINAD*. Therefore, the ADC is also evaluated according to the FOM_{DR} of eq. 4.1 where the input-referred *DR* is filled out. As such, the performance can be compared favorably to e.g. the fundamental limit of 1.6×10^{-20}J.

Interferer immunity

The evaluation of the interferer immunity concentrates on the setting where $R_{in} = 1k\Omega$ and the maximum wanted signal equals $3.5mV_{rms}$. The reason is that, especially in the case where the wanted signal is small, the interferers are likely to be larger. For the other settings, the results for nearby interferers are similar or even better. The allowable input level of far-off interferers, for these settings, is limited to the supply voltage because of reliability issues.

Table 9.5: *Measured performance of the feedback ADC (conventional parameters)*

bandwidth	0-1MHz		
sample rate	64MHz		
R_{in}	$100k\Omega$	$10k\Omega$	$1k\Omega$
full-scale, diff. input	$0.35V_{rms}$	$0.035V_{rms}$	$0.0035V_{rms}$
DR	71dB	64.5dB	49.5dB
peak-SNR	68.5dB	63dB	49dB
IM_3 at 500kHz	$<-66dB_c$	$<-63dB_c$	$<-61dB_c$
FOM_{SINAD} eq. 4.4	$7\times10^{-16}J$	$2\times10^{-15}J$	$6\times10^{-14}J$
overall FOM_{DR} eq. 4.1	$5\times10^{-18}J$		
$I_{quiescent}$	2.6mA @1.8V		
active area	$0.2mm^2$		
technology	$0.18\mu m$-CMOS, 1P, 5Al		

Aliasing limit on interferer immunity: Fig. 9.26 shows the result of an aliasing test (for $R_{in}=1k\Omega$) with an input of 64.37MHz at 0.02V and at 0.2V. The latter level is over 30dB higher than the full-scale level for wanted signals. The aliasing suppression equals 80dB here. For $R_{in}=10k\Omega$ it equals 90dB. These values are slightly lower than what is simulated (probably parasitic mixing is more important than aliasing). For $R_{in}=100k\Omega$ the alias suppression is over 100dB: the maximum input level is limited by the supply, the corresponding alias is hidden in the noise. As predicted in section 6.3 the aliasing suppression of this feedback ADC is significantly better than that of the feed forward reference ADC (even while the latter has a 5th-order loop filter). This is discussed in more detail in section 9.5.

Stable input range for interferers: In fig. 9.27 the measured stable input range is compared to the simulated graph[3]. Again, the simulation assumes small-signal operation and, by consequence, only gives a coarse prediction near full-scale. Transient simulations are much more accurate but are very time-consuming. Still, the slope of the curve is predicted well and also the absolute accuracy of the prediction is acceptable.

[3]Note that the simulated curve slightly differs from that in fig. 9.18. The latter was based on simulations of an ideal model. Here, simulations on the implemented loop filter are used and, moreover, the various coefficients have been slightly tuned to prevent overshoot in the adjacent channels.

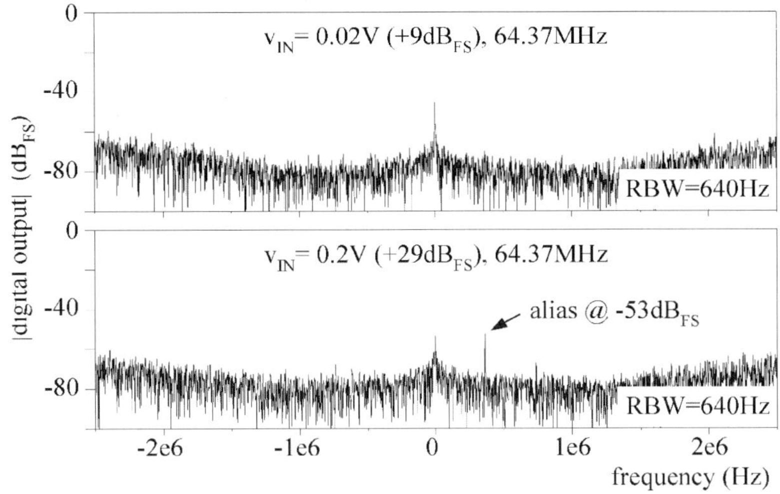

Figure 9.26: *Aliasing when applying an interferer at 64.37MHz at various input levels ($R_{in} = 1k\Omega$)*

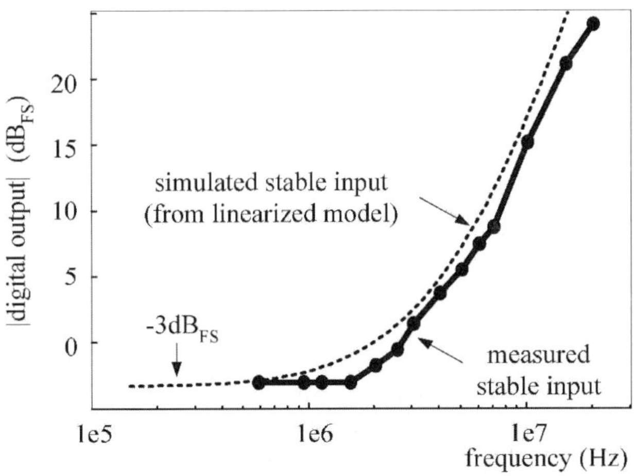

Figure 9.27: *Stable input range over frequency*

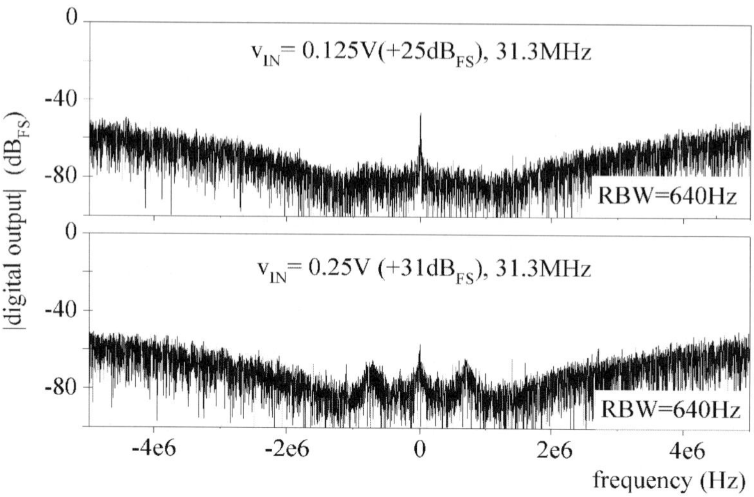

Figure 9.28: *Spurious responses when applying an interferer at 31.3MHz at various input levels ($R_{in}=1k\Omega$)*

Note that the curve is only depicted for positive frequencies. Contrary to fig. 9.14 for the feed forward reference ADC, here, the stable input range is more or less symmetrical around DC. This may be due to the order of the loop filter: the phase of the loop gain changes more gradually in the present 4th-order design.

Spurious responses due to interferers: An interferer at 31.3MHz is applied and its amplitude is increased until spurious components arise in the wanted channel (see fig. 9.28). For $R_{in}=1k\Omega$ an interferer of 0.125V may be applied at this frequency without affecting the *DR* listed in table 9.5. For the other settings, interferers at this frequency are only limited by the supply voltage. It can be concluded that the feedback ADC suffers much less from spurious responses than the feed forward ADC.

IM_3-distortion of interferers: The IM_3-distortion of the presented ADC strongly varies with the frequency of the input signals. Especially when adjacent interferers are applied, the internal signal swings may become large (see fig. 9.21) and cause distortion. Applying two adjacent interferers at frequencies f_1 and f_2 such that $2f_1 - f_2 = 500kHz$ yields the graph of fig. 9.29.a showing IM_3 as a function of f_1. The input level is adapted such that the interferers add up to the corresponding stable input range.

For $R_{in}=100k\Omega$ and $R_{in}=10k\Omega$, note the following:

- the IM_3-distortion is similar for both settings: although the input voltage is different in both cases, the input current is the same. By consequence, the internal signal swings and the resulting distortion is similar as well;

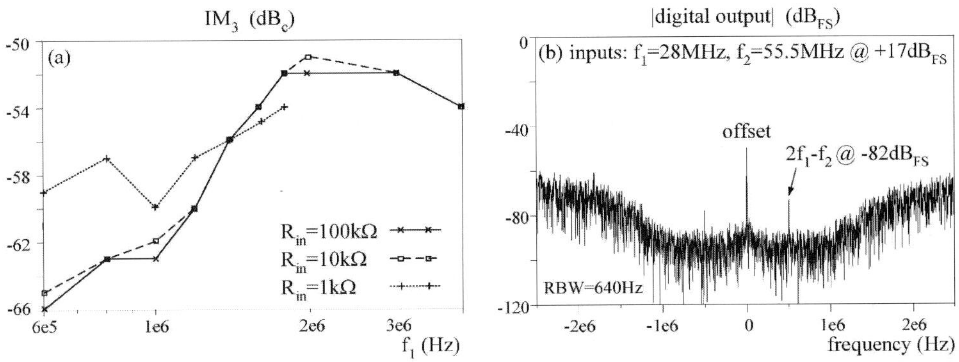

Figure 9.29: *IM_3 distortion as a function of the input frequency for adjacent interferers (a) and example for far-off interferers (R_{in} = 10kΩ) (b)*

- the IM_3-distortion changes with frequency. This is due to the frequency-dependent signal swing at the output of the 1st integrator. From fig. 9.21, this is the node with the largest signal swing (for adjacent interferers) and with the dominant distortion component.

The curves of fig. 9.29.a have been verified in simulation by applying an equivalent distortion source at the output of the first integrator. This model is also used to calculate the consequent limit on the allowable interferer level (see fig. 9.30). For $R_{in} = 1k\Omega$ the graph is different. This may be due to the following reasons:

- in this setting, the noise is significantly higher and partly adds to the distortion energy resulting in an inaccurate measurement. This is also the reason why IM_3 is not shown beyond 2MHz[4];

- the non-linearity of the input transconductors of the first stage becomes more important in this setting because of the low value for R_{in} (see eq. 4.12). By consequence, IM_3 is higher and less dependent on frequency than in the other modes.

It should be noted that the measured IM_3-distortion of adjacent interferers is significantly worse than that in the feed forward ADC. The main reason is the fact that, here, the entire signal is fed into the loop via nested feedback while in the feed forward implementation only the error signal is applied to the loop filter.

On the contrary, intermodulation is not a problem for *far-off interferers* (the internal signal swings are small at these frequencies). This is illustrated with the two-tone test of fig. 9.29.b: inputs at 28MHz and at 55.5MHz are applied at an amplitude of 0.5V for $R_{in} = 10k\Omega$ [5]. Note that this level is 20dB higher than the full-scale input level for wanted

[4] At these frequencies, the stable input range is smaller than in the wanted channel. Hence, for the IM_3-test the interferers must be applied with a smaller amplitude as well and the resulting IM_3-component gets hidden in the noise.

[5] This test is shown for $R_{in} = 10k\Omega$, because for $R_{in} = 1k\Omega$, intermodulation components are hidden in the noise.

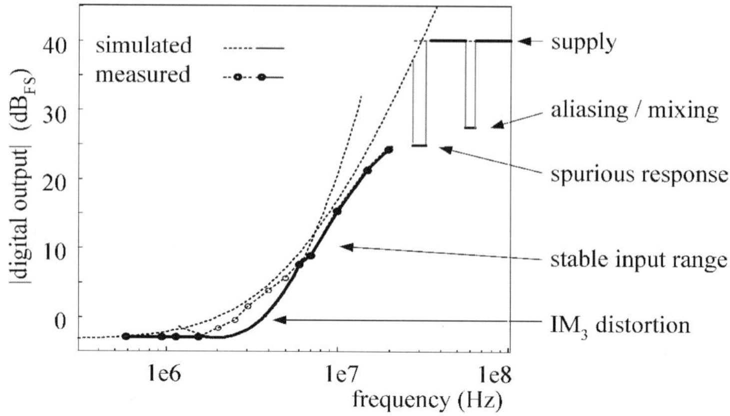

Figure 9.30: *Overall limit on the allowable interferer level over frequency ($R_{in}=1k\Omega$)*

signals for this gain setting. Even then, the resulting IM_3 component at 500kHz is at $-73dB_{FS}$. This is sufficiently low in order not to affect the *DR* listed in table 9.5.

The above evaluation is summarized in the graph of fig. 9.30 for $R_{in}=1k\Omega$. The limitation due to IM_3-distortion is depicted for a target IM_3-value of $-50dB_c$. This value is chosen in order to demonstrate that, contrary to the other conditioning ADCs evaluated in this chapter, *in the present design, IM_3-distortion may yield the dominant limitation on the allowable interferer level at adjacent frequencies. This depends on the target performance. In order to improve on this limit, further down-scaling of the unity-gain frequency of the first integrator is needed or more linear active stages must be designed. Both approaches will increase the power consumption of the ADC though. At higher frequencies, the allowable interferer level is set by the stable input range of the ADC and for even higher frequencies, the supply voltage is limiting the levels. Interferers applied near 32MHz must remain smaller than $+25dB_{FS}$ in view of spurious responses. Interferers applied near 64MHz are limited to $+28dB_{FS}$ in view of aliasing. For $R_{in}=10k\Omega$ and even more for $R_{in}=100k\Omega$, especially the supply limit becomes more restricting.*

9.4 FFB-ADC

A FFB-ADC is presented as a third implementation of the conditioning channel for the Bluetooth receiver. This design is an illustration of the theory developed in section 6.4. The ADC combines the advantages of the feed forward and the feedback topologies presented in the previous sections; i.e. it features both excellent power/performance and a high immunity to interferers.

9.4.1 Design

In an analogy to the previous section, the discussion on the ADC design covers some architectural choices (a.o. the choice of the loop filter and of the added filters), motivates the integration of programmable gain and analyzes some implementation related aspects.

ADC architecture

The FFB-ADC is constructed from a conventional ADC with a 4th-order, real, feed forward loop filter $H(s)$. (This ADC will be referred to as the "original ADC".) In fact, the 5th-order, complex loop filter of the ADC in section 9.2, could have been used, just as well. *However, a loop filter with real coefficients is preferred because it results in a lower complexity of the design, layout and evaluation. Moreover, the simulation time for a transient analysis reduces by more than a factor of two compared to a complex design. These arguments motivate the use of a "lower risk" loop filter with real coefficients in order to concentrate on the true innovation of this design; i.e. on the unrestricted filtering in the STF.* Furthermore, the choice for a design with real coefficients has the following consequences:

- for use in a quadrature receiver, two ADCs need to operate in parallel in order to digitize both the I and the Q signals;
- the effective over-sampling in a $\Sigma\Delta$ ADC with real coefficients is only half of that when using a complex loop filter. At this low over-sample factor, a 5th-order loop filter doesn't allow better noise shaping (because of stability) than a 4th-order filter. While in the complex designs, cross-coupling paths between quadrature signals were used to implement a resonance and to provide additional gain in the wanted channel. This is not possible in a design with real coefficients. For these two reasons, the *SQNR* of this ADC is only 74dB.

It is shown later on that, because of the filtering behavior of this ADC, programmable gain control can be added and, then, a much larger input-referred *DR* becomes available.

While the choice of the loop filter determines the SQNR and the NTF of the ADC, the STF can be designed independently, e.g. by adding complementary filters H_{LPF} *and* H_{HPF} *(see section 6.4).* In the presented design, the parallel configuration of fig. 6.19.a is used as this results in a favorable analog implementation. Likewise, for H_{LPF} and H_{HPF} a simple 1st-order implementation is used. Fig 9.31 shows the STF of the $4th$-order feed forward ADC without the complementary filters and the effect of adding the filters H_{LPF} and H_{HPF}. These filters are implemented with a $-3dB$-frequency of 3MHz. This choice results as a compromise between:

- good attenuation of adjacent interferers and of the overshoot that is present in the STF of the original feed forward ADC;
- limited droop and group delay variation in the wanted channel.

The latter requirements are application dependent and are outside of the present scope. In general, the STF of the FFB-ADC is optimized by choosing the appropriate charac-

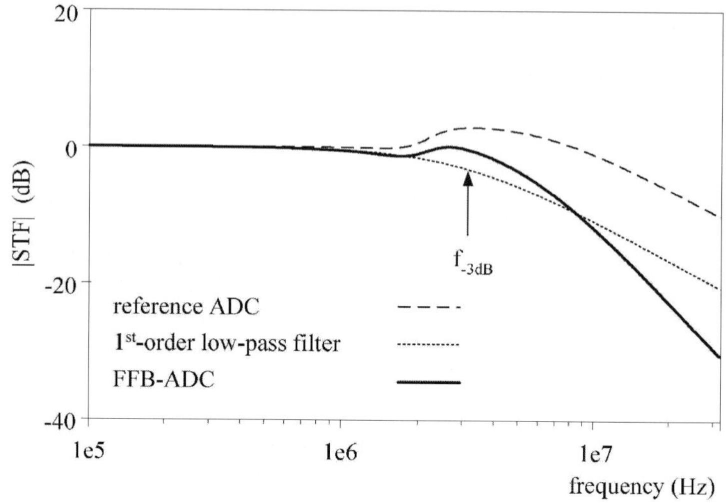

Figure 9.31: *STF of the FFB-ADC with unrestricted filtering*

teristic for H_{LPF} and H_{HPF}. On the contrary, the loop filter $H(s)$ should not be modified compared to that of the original ADC as it has been designed for optimal noise shaping.

Integration of programmable gain

As argued for the feedback ADC on page 159, a filtering STF is optimally exploited in combination with programmable gain control of the input signal. Since the ADC is highly immune to interferers, its input range can be scaled according to the strength of the incoming wanted signal. As such, the input-referred *DR* is extended significantly. Moreover, the merge of programmable gain and filtering into a ΣΔ ADC yields a true equivalent of a conventional conditioning channel.

The gain programming is implemented in the same way as was done for the feedback ADC: the input resistance of the ADC is switched between $1k\Omega$, $10k\Omega$ and $100k\Omega$ in order to, respectively, accommodate wanted signals up to 5mV, 50mV and 500mV of amplitude. In addition, in the present design, the quiescent current of the first stage is adapted along with the input resistance[6]. This is motivated further on as one of the implementation aspects.

Implementation aspects

Fig. 9.32 shows a block diagram of the implementation. Compared to fig. 6.19.a, the first stage $H_1(s)$ of the loop filter $H(s)$ is shifted in front of the summation point and is

[6]The scaling of the quiescent current can be implemented in the feedback design as well and, likewise, would improve the average power/performance ratio.

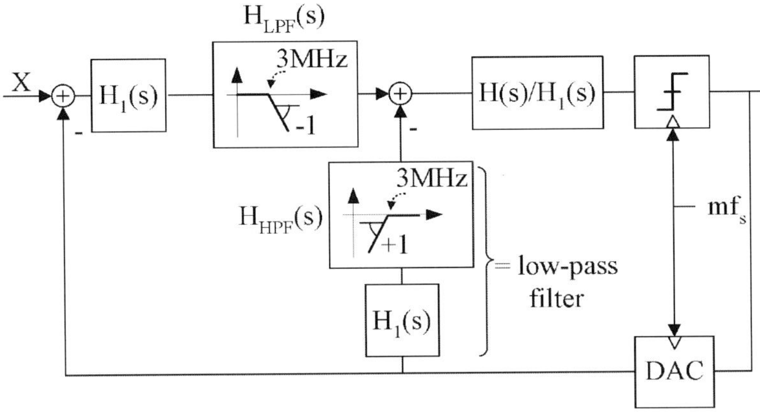

Figure 9.32: *Block diagram of the implemented FFB-ADC*

duplicated in the parallel feedback path. Shifting $H_1(s)$ before H_{LPF} and H_{HPF} yields the advantage of an easier implementation for these filters:

- H_{LPF} is in between two active stages (acting as buffers) and a passive implementation becomes possible;
- the series connection of the integrator $H_1(s)$ and the 1st-order high-pass filter H_{HPF} can be replaced by a low-pass filter, resulting in an easier implementation.

Fig. 9.33 shows the block diagram of the actual implementation. The 4th-order loop filter is implemented in a similar way as in the previous designs. The resonance around the 4th integrator provides some additional gain at the edge of the wanted channel. The combination of the resistor R_f and the subsequent capacitor C_1 constitutes H_{LPF}. (For simplicity, a floating capacitor of value $C_1/2$ is drawn. The actual implementation uses two capacitors of value C_1 that are referenced to ground.)

In the parallel feedback path, the equivalent low-pass filter is built around OTA g_{fb}. The combination of R_{fb} and C_1 realizes $H_1(s)$ and the combination of R_f and C_1 realizes H_{HPF}. Note that, here, an active implementation is preferred over a passive one: the active stage supplies the signal-dependent current to the integration capacitor such that the DAC reference has a constant load. A passive implementation would pollute the DAC reference that is being shared with other circuits, with a signal-dependent load.

Finally, in order to perform the summation of the parallel path in the current domain, two additional transconductors fb_1 and fb_2 are needed. Their quiescent current of $10\mu A$ is negligible.

Internal signal swings: When discussing the feedback topology of section 9.3, the signal swing at the output of the first integrator was identified as an important problem. It is the dominant source of distortion and limits the unity-gain frequency -and thus the gain- of the first integrator. Here, the addition of the filters also affects the signal swing at this node.

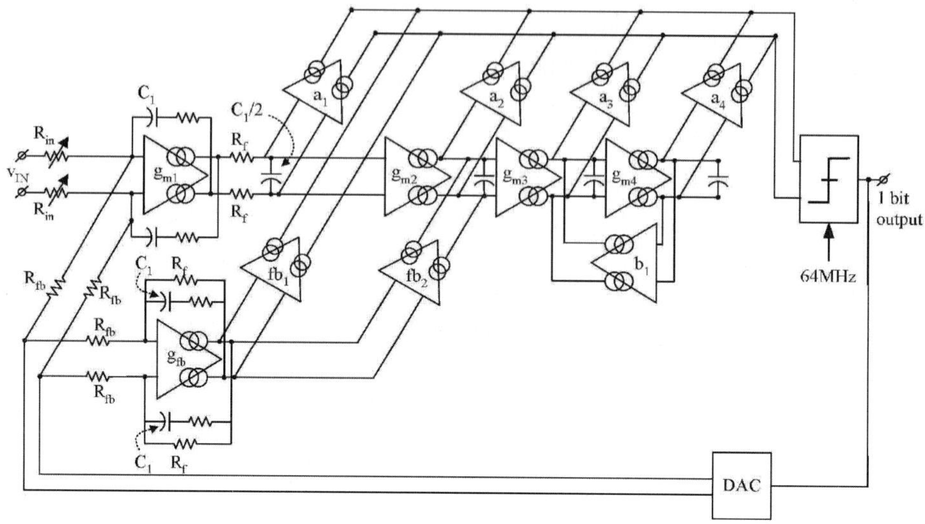

Figure 9.33: 4th*-order ΣΔ ADC with unrestricted filtering*

Compared to the feed forward ADC of section 9.2, the signal swing has increased in the wanted channel (see fig. 9.34). However, the maximum signal swing remains comparable. In addition, the unity-gain frequency of the first integrator (f_{ug}=*2.9MHz*) is about equal to that of feed forward ADC of section 9.2 (f_{ug}=*2.8MHz*).

Compared to the feedback implementation of section 9.3, though, the signal remains significantly smaller, even while the unity-gain frequency of the first integrator is 1.5 times larger here.

Noise and distortion of the added filters: The major noise contributions of the design are shown in fig. 9.35 that is discussed further on. Since H_1 has been shifted in front, the noise and distortion of H_{LPF} is counteracted by the preceding gain. The same is true is for OTA g_{fb} in the feedback path: when referring to the input, its contribution to noise and distortion is suppressed by H_1. This OTA is a scaled copy of g_{m1}: while g_{m1} consumes $500\mu A$, g_{fb} can be biased at only $50\mu A$ because of the above reasons. For use in section 6.4.3 it is mentioned that the implementation factor of the FFB-ADC, as compared to the original ADC, equals $F \cong 1.1$.

It can be concluded that the added filters hardly increase the overall noise, distortion or current consumption of the original ADC. The basic reason is the fact that the added filters are put inside of a closed-loop system. This is a key difference compared to the conventional cascade of an analog filter and an ADC. There, the filter would be in the signal path and, consequently, its noise and distortion would contribute.

Programmable quiescent current for the first stage: As in the previous ADCs, the first stage of the loop filter contributes dominantly to the overall circuit noise. There-

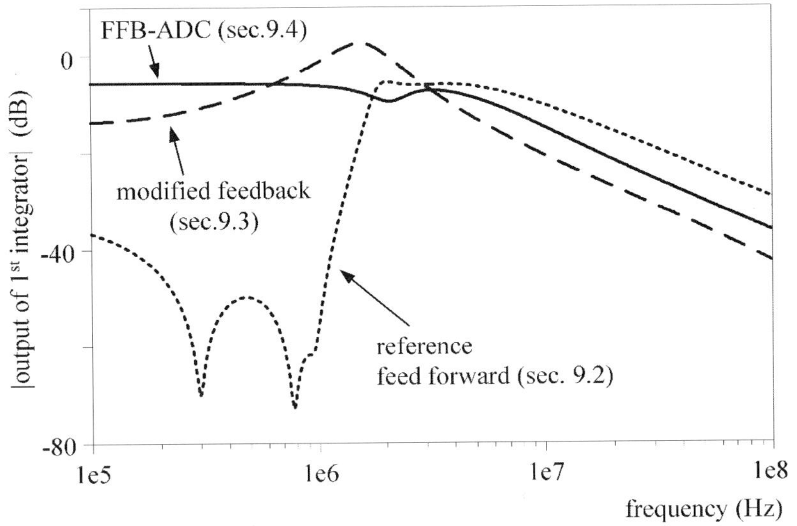

Figure 9.34: *Transfer function from the input of the ADC to the output of the first integrator based on a linearized model*

fore, its nominal quiescent current is set at $500\mu A$ (just as in the input stage of previous ADCs). A small improvement of the average consumption is implemented by adapting this current along with the programming of R_{in} to the magnitude of the wanted signal. As such, the quiescent current of the first stage is reduced to $200\mu A$ when R_{in} is switched to $100k\Omega$ because a large input signal is present.

Fig 9.35 shows the simulated, input-referred noise-density and the major contributions. For $R_{in}=1k\Omega$ (and for $R_{in}=10k\Omega$) the first integrator is dominant. For $R_{in}=100k\Omega$ the contribution of g_{m2} and of g_{fb} become important because the preceding gain drops (the gain of the input integrator is inversely proportional to R_{in}). In general terms, the effect of the programming of R_{in} on the various noise contributions is similar to what has been discussed for the feedback ADC on page 159.

Mismatch between H_{LPF} and H_{HPF}: A mismatch between the time-constants of the added filters H_{LPF} and H_{HPF} would violate the complementarity posted in eq. 6.19. Theoretically, this affects the stability of the loop and an additional phase margin must then be taken into account during the design phase. For common mismatch values, though, this issue is not at all restricting. For example, it can be calculated that a 5%-mismatch on the time-constants causes a phase shift of less than 0.3 degrees at half the sample rate.

Spread on H_{LPF} and H_{HPF}: Spread on the target $-3dB$-frequency of the filters results in spread on the STF of the ADC and thus on the filtering of the interferers. If this spread cannot be tolerated, this problem must be dealt with as in any continuous-time filter; i.e. either the time-constants are calibrated or a margin on the filtering is implemented.

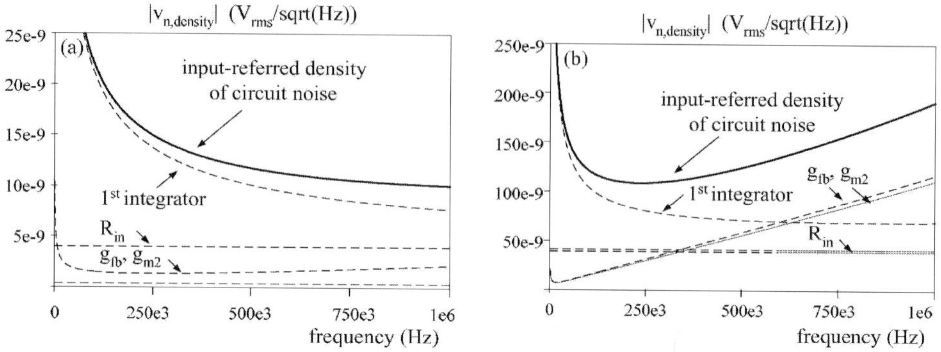

Figure 9.35: *Input-referred circuit noise for $R_{in}=1k\Omega$ (a) and $R_{in}=100k\Omega$ (b)*

9.4.2 Evaluation

Similar to the ADCs of section 9.2 and 9.3, this design is embedded in a test-IC including a bandgap reference and on-chip clock generation. The evaluation and the discussion hereon are conducted in a similar way as well.

Conventional performance parameters

Fig 9.36 shows the measured *SNR* as a function of the input voltage for the various values of R_{in}. Adapting the value of R_{in} depending on the magnitude of the wanted channel results in an input-referred *DR* of 89dB. In addition, the output-referred peak-*SNR* remains moderate. It is discussed further on (section 9.5) that this is beneficial on a system level. As for the feedback ADC, the difference in peak-*SNR* between the various settings is due to a different relative contribution of circuit noise and quantization noise.

Fig. 9.37 shows the output spectrum of an IM_3-test in the wanted channel. For $R_{in}=100k\Omega$, IM_2- and HD_2-components are visible. These have been traced back to the generator[7]. In addition, some shaping of quantization noise is visible within the wanted channel. For $R_{in}=1k\Omega$, circuit noise is by far dominant and the noise-density corresponds to the curve simulated in fig. 9.35.a. In general, $IM_3 < -60dB_c$ for $R_{in}=100k\Omega$ and $R_{in}=10k\Omega$. For $R_{in}=1k\Omega$, it is below $-50dB_c$. Qualitatively, this increase is predicted from eq. 4.12 and considering the values for v_{in}, R_{in} and g_m. Quantitatively, the difference deviates from what is predicted by eq. 4.12. Compared to the IM_3-distortion for $R_{in}=100k\Omega$, eq. 4.12 predicts 4dB lower and 16dB higher distortion for $R_{in}=10k\Omega$ and $R_{in}=1k\Omega$ respectively. A possible explanation may be in the fact that the input transistors operate in weak (instead of strong) inversion or in the fact that other distortion sources contribute as well.

[7] For most other test, generator induced spurious are suppressed by calibration before measuring.

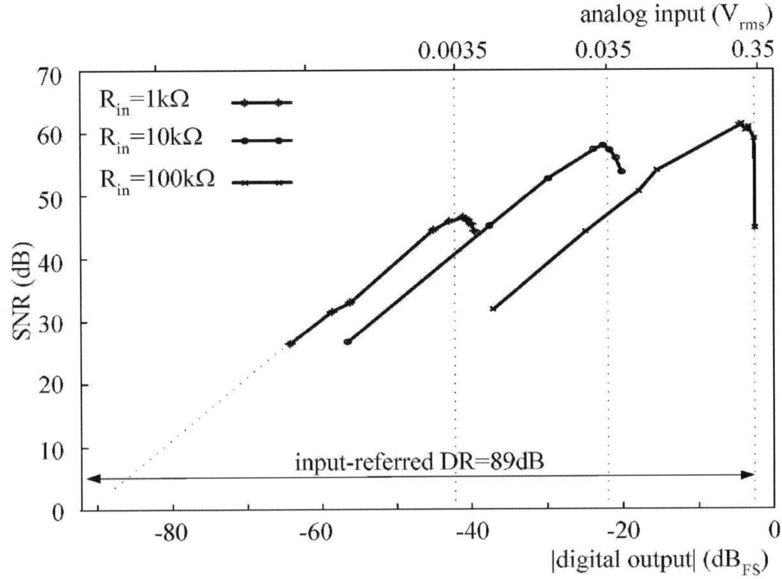

Figure 9.36: *SNR versus input swing for the three settings of R_{in}*

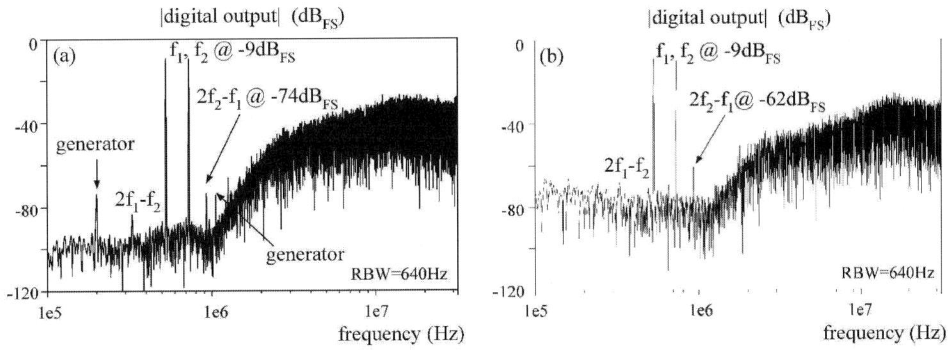

Figure 9.37: *Two-tone test in the wanted channel evaluating IM_3-distortion for $R_{in} = 100k\Omega$(a) and for $R_{in} = 1k\Omega$ (b)*

In the wanted channel, the measured IM_3-distortion is more or less independent of frequency. Hence, most likely, it is due to a non-linearity in the input stage or in the feedback path.

Table 9.6 gives on overview of the measured conventional performance parameters. In addition, it lists the achieved FOM_{SINAD} according to eq. 4.4. Again, this FOM is not suitable for the evaluation of conditioning ADCs because these limit the *SINAD* towards the output. Therefore, eq. 4.1 is evaluated as well. It benchmarks the ADC according to the input-referred *DR* it provides. This table lists the performance of one ADC with real

Table 9.6: *Measured performance of one real, FFB-ADC (conventional parameters)*

bandwidth	0-1MHz		
sample rate	64MHz		
Nyquist sample rate	2MHz		
R_{in}	$100k\Omega$	$10k\Omega$	$1k\Omega$
full-scale, diff. input	$0.35V_{rms}$	$0.035V_{rms}$	$0.0035V_{rms}$
DR	65dB	59dB	49dB
peak-SNR	59dB	57dB	46dB
IM_3 (in wanted channel)	$<-60dB_c$	$<-60dB_c$	$<-50dB_c$
$I_{quiescent}$	0.85mA	0.85mA	1.15mA
FOM_{SINAD} eq. 4.4	$2 \times 10^{-15}J$	$3 \times 10^{-15}J$	$5 \times 10^{-14}J$
overall FOM_{DR} eq. 4.1	$2.6 \times 10^{-18}J$		
active area	$0.14mm^2$		
technology	$0.18\mu m$-CMOS, 1P, 5Al		

coefficients. For comparison with the complex implementations of the previous sections, the performance, area and power consumption of two parallel ADCs should be considered. Then, in the quadrature configuration, a 3dB higher *DR* and peak-*SNR* are achieved and area and power consumption double.

Interferer immunity

Fig. 9.38 illustrates the operation of the presented FFB-ADC. A very small wanted input is applied together with two much larger, far-off interferers. The latter are applied at an input level corresponding to half the stable input for that frequency. At the output, the wanted signal appears near the full-scale output level. Relatively, the interferers, have been attenuated. This demonstrates the filtering STF of the presented ADC. In addition, this measurement also proves the linearity with respect to far-off interferers. Even though the interferers are applied at an input level that is significantly higher than the wanted channel (f_2 is applied 15dB higher, f_3 is applied 24dB higher), at the output, their IM_3-distortion is 55dB below the wanted channel.

The measurement of fig. 9.38 gives a first impression of the interferer immunity of this ADC. A systematic evaluation of the various limitations on the allowable interferer level is presented below. Afterwards, the most restricting limitations are summarized.

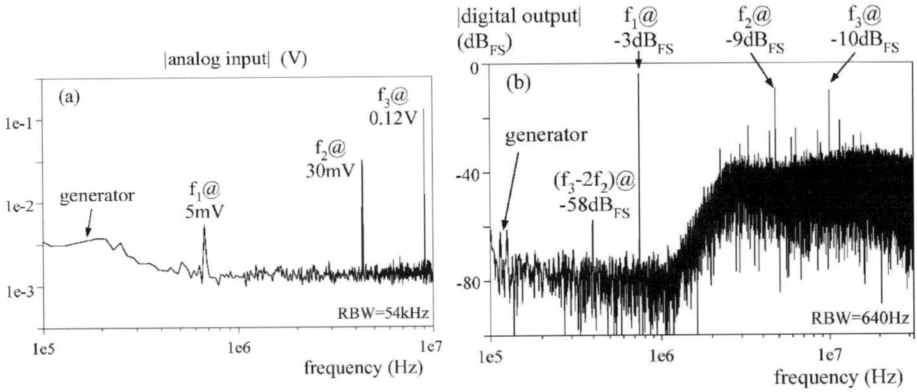

Figure 9.38: *Tri-tone input (a) and output spectrum (b) demonstrating filtering and linearity ($f_1 = 700kHz$, $f_2 = 4.8MHz$ and $f_3 = 10MHz$; $R_{in} = 1k\Omega$)*

Aliasing limit on interferer immunity: Alias suppression of 69dB, 67dB, and 76dB is measured for R_{in} equaling $100k\Omega$, $10k\Omega$ and $1k\Omega$ respectively. These values are lower than what is expected from eq. 6.3. Probably, parasitic mixing with the clock frequency is dominant over aliasing. This hypothesis, would also explain the difference in the measured values.

The alias suppression is better than that of the feed forward ADC of section 9.2 (60dB of suppression is achieved there) but worse than that of the feedback ADC of section 9.3 (achieving over 80dB of alias suppression). This is explained by the difference in the order of L_0 and of L_1. This is discussed in more detail in section 9.5.

Stable input range for interferers: Fig. 9.39 compares the measured stable input to the simulated level based on a linearized small-signal model for $R_{in} = 1k\Omega$. For adjacent frequencies, the prediction from the model may deviate from the measurement by a few decibels. For far-off frequencies, the prediction of the absolute value and of the slope of the curve is very good. Measurements over several samples and also for $R_{in} = 10k\Omega$, show that the maximum deviation between the simulated and the measured overload level is about 2dB.

Spurious responses due to interferers: In this design, the order of L_0 equals 2 near half the sample frequency. Compared to the feed forward ADC of section 9.2 (with L_0 of order 1 in this frequency range), the interferer is attenuated stronger before being applied to the quantizer. Since the quantizer represents a "strong non-linearity", this results in a major improvement in terms of spurious responses. Following a similar reasoning, this ADC must be somewhat less robust, in this respect, than the feedback ADC of section 9.3 having L_0 of order 3 near half the sample frequency. This is demonstrated by the measurement in fig. 9.40. For $R_{in} = 1k\Omega$, interferers up to 100mV can be applied without affecting the *DR* in the wanted channel. Note, the full-scale input for wanted

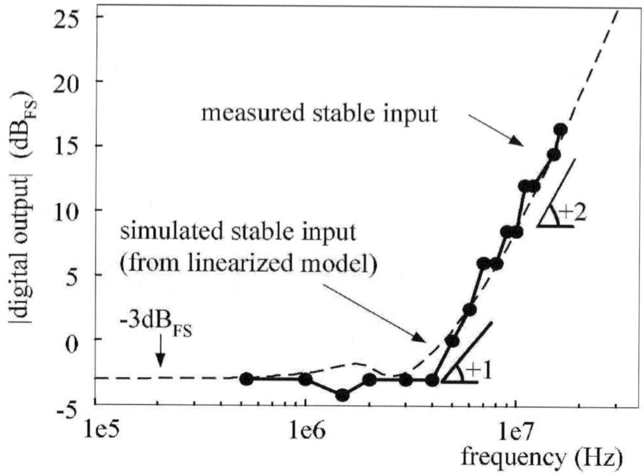

Figure 9.39: *Stable input range over frequency*

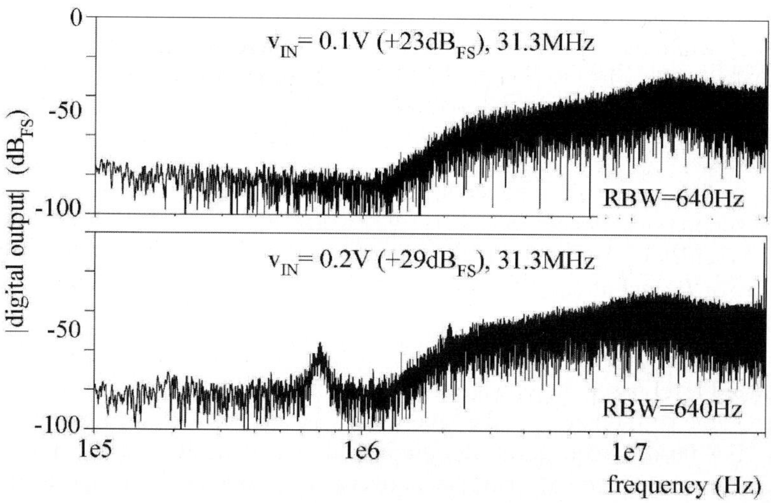

Figure 9.40: *Spurious responses when applying an interferer at 31.3MHz at various input levels ($R_{in}=1k\Omega$)*

channels is only 5mV in this setting. For $R_{in}=10k\Omega$ and $R_{in}=100k\Omega$, the allowable interferer level near 32MHz is limited by the supply voltage.

IM_3-distortion of interferers: For $R_{in}=1k\Omega$, fig. 9.41 shows IM_3-distortion over frequency. In this measurement, interferers at frequency f_1 and frequency f_2 are applied at an input level corresponding to half the stable input range for that frequency (see fig. 9.39). These frequencies are chosen such that $2f_1-f_2$ falls in the wanted

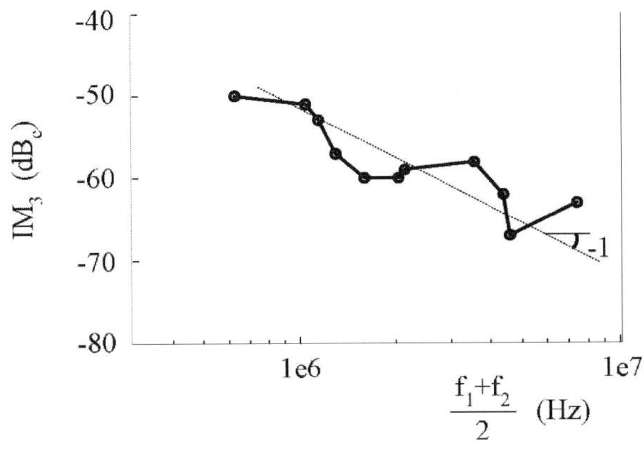

Figure 9.41: *Measured IM_3-distortion of interferers for $R_{in}=1k\Omega$*

channel. Within the measured frequency range, IM_3-distortion decreases as a function of frequency while the input level of the interferers increases according to the curve of fig. 9.39. Hence, the distorting node may be inside of the loop. This is not investigated further because, nor for adjacent interferers, nor for far-off interferers, IM_3-distortion is a restricting limit on the allowable interferer level. This becomes clear from the discussion on fig. 9.42 below. The same conclusion holds for $R_{in}=10k\Omega$ and $R_{in}=100k\Omega$.

Fig. 9.42 gives a summary of the limitations on the allowable interferer level over frequency for $R_{in}=1k\Omega$. For most frequencies, the stable input range causes the restricting limit. Near 32MHz and near 64MHz the interferers must be smaller than $+23dB_{FS}$ and $+24dB_{FS}$ respectively to keep spurious responses and alias components low enough. IM_3-distortion is not a restricting limit for this ADC For $R_{in}=10k\Omega$ and $R_{in}=100k\Omega$, the graph is similar except for the far-off interferers being limited by the supply voltage.

9.5 Evaluation of the channels

In this chapter, three solutions for the signal conditioning channel in a highly-digitized Bluetooth receiver have been presented. First, in section 9.5.1, the presented ADCs are briefly benchmarked to published designs with a similar bandwidth. Next, in section 9.5.2, the ADCs are compared against each other. Since all have been implemented in the same technology and -to a large extent- reuse the same circuits, this is a fair comparison. Finally, the presented concept of digitizing the channel by integrating filtering and programmable gain into the ADC, is evaluated by comparison with highly-analog channels and with channels using "conventional digitization".

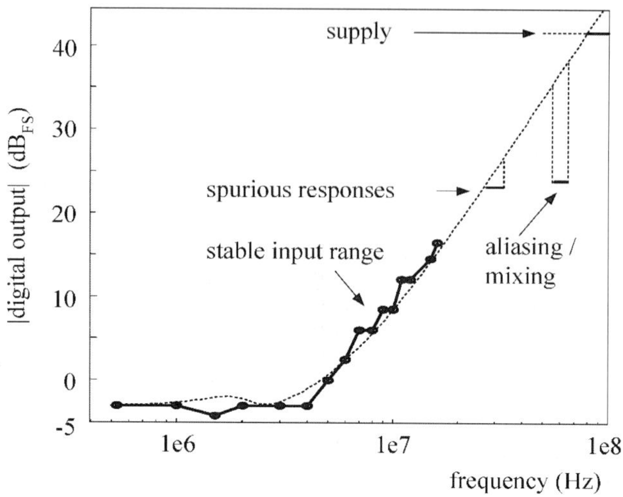

Figure 9.42: *Overall limit on the allowable interferer level over frequency*

9.5.1 Benchmark with published ADCs

Only for the reference ADC of section 9.2, this benchmark is straightforward. It can be evaluated using the FOM_{SINAD} of eq. 4.4 and compared to other ADCs, a.o. [33], [123], [124], [39], [28] and [125], published for this application area and listed in table A.1. Together with [28], it achieves the best FOM among the converters targeting a bandwidth of 1MHz.

The conditioning ADCs of sections 9.3 and 9.4 cannot be benchmarked in this manner because they target to reduce the *SINAD* towards the output. Instead, the FOM_{DR} of eq. 4.1 can be used to evaluate their input-referred *DR*. In this aspect they outperform "conventional ADCs" by two orders of magnitude, typically; i.e. FOM_{DR} is of the order of 10^{-19}J for the conditioning ADCs while it is $\sim 10^{-16}$J for state-of-the-art "conventional ADCs". This demonstrates that *a major power saving is possible when targeting optimization of the overall channel instead of optimizing only building blocks.*

9.5.2 Comparison of the presented ADCs

The ADCs of sections 9.2, 9.3 and 9.4 are compared to each other. For simplicity, they are referred to, here, as the feed forward ADC, the feedback ADC and the FFB-ADC respectively. Table 9.7 gives an overview of the conventional performance parameters for the three ADCs. The figures in the last column refer to the quadrature configuration of two FFB-ADCs. Since three settings are possible for the operation of the feedback and the FFB-ADCs, their performance can be summarized using the combination of the input-referred *DR* and the *SNR* at the output. The discussion on this table distinguishes between conventional performance parameters and interferer immunity.

Table 9.7: *Comparison of the ADCs in terms of conventional parameters*

	feed forward ADC (sec. 9.2)	feedback ADC (sec. 9.3)	2 filtering ADCs (sec. 9.4)
active area	$0.22mm^2$	$0.2mm^2$	$0.27mm^2$
power consumption	4.4mW	4.7mW	<4.1mW
input-referred *DR*	76dB	89.5dB	92dB
peak-*SNR* at output	75.5dB	68.5dB	62dB
FOM_{DR} eq. 4.1	$1 \times 10^{-16} J$	$5.3 \times 10^{-18} J$	$2.6 \times 10^{-18} J$
accuracy of references	0.01%	0.2%	0.2%

Comparison of conventional performance parameters: All solutions have a comparable power consumption and occupy a similar chip area. Two observations stand out.

First, *the solutions differ in the fact that both the feedback ADC and the FFB-ADC have a significantly larger input-referred DR and a much moderate peak-SNR at the output than the feed forward reference ADC. This advantage is obtained from the architectural innovation of integrating explicit signal conditioning into the ADC.* The conditioning channel as a whole becomes more efficient in terms of power/performance:

- *FOM_{DR} of the ADC improves by two orders of magnitude:* for a fixed power consumption and maximum signal level, a larger input-referred *DR* is achieved. IM_3 distortion is lower though. In fact, all performance figures evolve in a direction that matches the channel requirements: a large *DR* is highly desirable, on the contrary, IM_3-distortion of the wanted channel can be as high as $-18dBc$ without affecting the bit error rate.

- *A power and area reduction for the analog reference circuits:* since the DAC in the feedback path of the ADC can have a lower peak-*SNR*, the reference circuits for the DAC can have about 20 times lower accuracy as well. The required accuracy[8] is listed in the last row of table 9.7. This is the accuracy needed to maintain the full *DR* for $R_{in} = 1k\Omega$ and listed in tables 9.5 and 9.6.

[8] These numbers only give a first-order indication. The allowed "noise" depends on the frequency spectrum, the stochastic distribution and on the shape of the DAC output pulses [126].

- *Some power and area reduction in the following blocks of the channel:* the decimation filter can be dimensioned to a smaller peak-*SNR*, reducing the value of ENOB in eq. 4.19. The improvement in peak-*SNR*, as listed in table 9.7, is 7dB for the feedback ADC and 13.5dB for the FFB-ADC. From eq. 4.19, the power consumption of the decimation filter may then reduce by 20% up to 40%[9]. The bit rate in between digital blocks (e.g. over an inter-die interface) is reduced as well. This saving can be derived from eq. 7.4 but is minor here.

In general, the smaller peak-*SNR* at the output may also allow a lower sample rate for the ΣΔ ADC (here, only the order of the loop filter has been changed). This would further simplify the clock generation (next to the milder accuracy requirement) and reduce the input sample rate of the decimation filter (next to the lower ENOB requirement).

Second, *the FFB-ADC achieves a 2-times better FOM than the feedback ADC.* This reflects the discussion in section 9.3 on the difficult balancing of power/performance. In the feedback ADC, noise and distortion are balanced by tuning coefficients and tweaking quiescent currents of internal stages. On the contrary, the FFB-ADC is designed from a topology that -inherently- achieves a better power/performance because of the overall feedback.

Comparison of interferer immunity: This comparison is conducted based on the graphs in fig. 9.43. For clarification of the graphs, first note the following.

On the y-axis, the antenna-referred signal power is plotted (instead of the input signal to the ADC) in order to easily link with specifications from the Bluetooth standard [112]. For example, the maximum specified input power equals $-20dBm$. The gain of the receiver front-end must then be designed such that this signal is applied at the full-scale input of the ADC.

The dotted line represents the envelope curve of the specified maximum input signals over frequency. Note, the standard does not require that the receiver simultaneously accommodates these maximum inputs.

The solid line indicates the allowable input level over frequency as introduced in fig. 9.16 (positive frequencies only) and fig. 9.30. At the bottom end of the scale, the level of the input referred noise of the ADC is indicated.

Hence, the interferer immunity of the feedback ADC of section 9.3 is -in first order- comparable to that of the FFB-ADC of section 9.4. Therefore, it is left out of the first part of the discussion, comparing the feed forward reference ADC to the FFB-ADC. Afterwards, the feedback ADC is briefly compared to the FFB-ADC.

The feed forward reference ADC of section 9.2 can accommodate all specified interferer levels and provides sufficient *DR*. (A receiver sensitivity of $-70dBm$ is required.

[9] Depending on the required interferer immunity, more PGA-settings can be applied and the peak-*SNR* of the ADC can be further reduced. This leads to a more significant saving in the decimation filter. The automatic gain control becomes more difficult though.

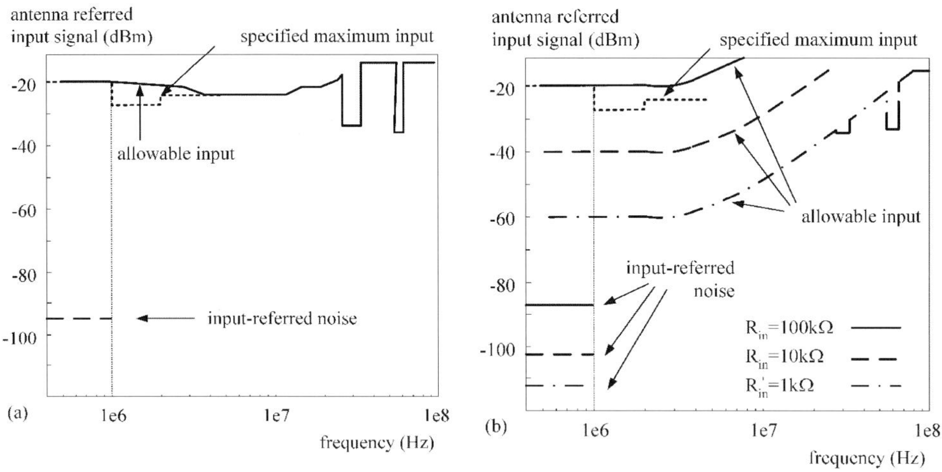

Figure 9.43: *Comparison of the dynamic range and interferer immunity (referred to the input power at the antenna) of the feed forward reference ADC (a) and of the quadrature configuration of two FFB-ADCs (b)*

At that input level, 18dB of *SNR* is needed for the digital demodulation. In addition, 8dB margin is foreseen in view of the front-end noise). As such, *the conventional feed forward ADC enables Bluetooth certification of the receiver of fig. 9.2, even while no analog filters or VGA are used. However, the margin on the interferer immunity is tight.* In addition, the de-facto sensitivity of published Bluetooth receivers is around *−82dBm* (a.o. [127], [115], [116], [120], etc.) or lower (*−88dBm* is reported in [128]). In order to be competitive, the present receiver needs more *DR* in the ADC or it needs preceding analog conditioning.

On the contrary, *for the same peak power consumption, the FFB-ADC does enable a Bluetooth receiver achieving the de-facto sensitivity goal and featuring a much more robust operation in the field:*

- For $R_{in} = 100k\Omega$, the basic Bluetooth specifications (with respect to noise and interferer tests) are met. Still, little margin on the sensitivity is available.

- Switching R_{in} to a lower value improves the sensitivity (the antenna-referred noise of the ADC is $-112dBm$ for $R_{in} = 1k\Omega$).

- The inherent interferer immunity considerably improves the operation in the field (with other or more combinations of interferers than those specified in the Bluetooth standard) and the overall robustness of the receiver.

Unfortunately, the latter advantage -although extremely important- is difficult to quantify uniquely. That would require a multitude of tests in order to evaluate all common combinations of interferers at various amplitude levels of the wanted channel.

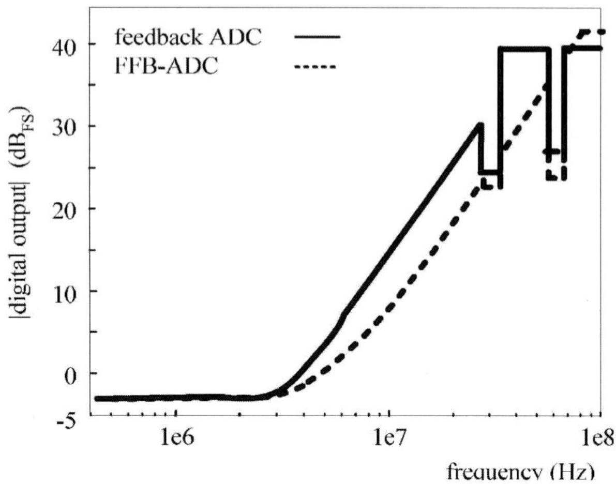

Figure 9.44: *Allowable interferer level of the feedback ADC of section 9.3 compared to that of the FFB-ADC of section 9.4*

Alternatively, the concept of the conditioning ΣΔ ADC could be used to lower the power consumption of the receiver while keeping the performance equal to that of the conventional feed forward ADC.

Finally, *the interferer immunity of the feedback ADC of section 9.3 is compared to that of the FFB-ADC of section 9.4.* This comparison is visualized in fig. 9.44. The following is observed:

- For far-off interferers, the feedback ADC is more immune. This is especially due to its larger stable input range at these frequencies and to a better alias-suppression. Still, for many standards and in many receiver topologies, the immunity achieved in the FFB-ADC may be well sufficient.
- For adjacent channels, both ADCs achieve a similar degree of immunity[10]. The feedback ADC attenuates adjacent channels more but its distortion limits the allowable input to a level comparable to that of the FFB-ADC.

For the feedback ADC, the latter limit can only be improved at the expense of a higher power consumption (see discussion on page 170). In the FFB-ADC, the allowable adjacent interferer level is limited by the stable input range. It can be improved by implementing a smaller f_{-3dB} for the added filters or by changing the order of the filter. This may have a noise penalty (the noise of a.o. the second stage of the loop filter becomes more important) or a distortion penalty (the internal signal swings increase) respectively. Still, *fundamentally, more degrees-of-freedom are available for the optimization of the FFB-ADC* and therefore, the power penalty can be kept minimal.

[10]This is true for the current example of an IM_3-target of $-50dB_c$ and with the current implementation parameters.

Table 9.8: *Comparison of the ADCs for operation in a wireless system*

	feed forward ADC (sec. 9.2)	feedback ADC (sec. 9.3)	FFB-ADC (sec. 9.4)
power/performance [a]	+	++	+++
interferer immunity	+	++	++
opportunities for further optimization	?	+	++

[a] Assuming only a moderate *SNR* is needed for demodulation, a power reduction is obtained by reducing the peak-*SNR* early on in the receive chain, i.e. at the output of the ADC. In other words, in these systems a moderate peak-*SNR* (in combination with a large input *DR*) is favorable.

These degrees-of-freedom originate from the addition of the extra filters. This brings us to another limitation of the feedback ADC; i.e. the NTF and STF cannot be optimized independently. Designing an optimal NTF results in a fixed value for f_{-3dB} of the STF. In general, if more over-sampling is applied and optimal noise shaping is pursued, the loop gain remains high over a wider frequency range. Consequently, f_{-3dB} of the STF increases.

The discussion is summarized in table 9.8. The feed forward ADC achieves a very competitive power/performance ratio comparing to state-of-the-art. In addition, it is sufficiently immune to interferers in order to enable a Bluetooth receiver without analog signal conditioning. It remains an attractive solution for systems requiring a high *SNR* for the digital processing. As an example, ADSL -using a more complex modulation scheme- could be considered.

Most wireless communication systems, though, use a modulation scheme that needs less *SNR* for demodulation. This is also the case for the present Bluetooth receiver. Here, the feedback and the FFB-ADC enable a much better sensitivity and higher robustness. Especially, the FFB-ADC is attractive. The present implementation performs slightly better than the feedback ADC. More important, more degrees-of-freedom are available for further improvements or for adaptation to other interferer spectra.

9.5.3 Benchmark with published Bluetooth conditioning channels

Publications on Bluetooth receivers hardly present data on the baseband part of the signal conditioning channel separately. Hence, the RF part is included in the present benchmark. The RF part of the low IF receiver (the receiver of fig. 9.2) consists of an LNA and a passive mixer only. It achieves $-71dBm$ of sensitivity while the peak power consumption is 32mW [121]. Table 9.9 compares this receiver to an analog Bluetooth receiver [127]

Table 9.9: *Benchmark of the Bluetooth receiver*

	an. con. + an. demod. [127]	an. con. + dig. demod. [120]	dig. con. + dig. demod. [121]+ ADC of sec. 9.2
sensitivity	-82dBm	-83dBm	-71dBm
max. input	0dBm	-5dBm	-19dBm[a]
P	75mW	58mW	32mW
area	$4mm^2$	$?mm^2$	$3.5mm^2$
technology	$0.18\mu m$ CMOS	$0.13\mu m$ CMOS	$0.18\mu m$ CMOS

[a]The maximum input power is smaller than expected. Probably, this is due to a problem in the matching network at RF.

and to a receiver using analog signal conditioning but digital demodulation [120]. (The latter receiver uses various sections of discrete-time filtering and a switched-capacitor ΣΔ ADC.) To the author's knowledge, these receivers are among the best reported. *Note that, as opposed to the various solutions presented in this chapter, all reported Bluetooth receivers need analog signal conditioning.*

A comparison of power/performance between the three receivers of table 9.9 is not straightforward because of the difference in sensitivity. It requires extrapolation of the power consumption of the receiver of the last column (i.e. of fig. 9.2) to a sensitivity of $-83dBm$. Probably, the highest power increase is needed in the LNA and in the mixer (likely the passive mixer must be replaced by an active solution to meet the performance target). The PLL and the VCO can remain unaltered because their performance is believed to be good enough. Most important, the ADC of section 9.2 can be replaced by the ADC of section 9.3 or, preferably, by the ADC of section 9.4. This can be done without any power penalty at all while the required *DR* is achieved with a margin of about 10dB. The decimation filter that follows, becomes easier than in the present channel. On the other hand, an automatic gain control algorithm must be implemented.

Although extrapolation of the power consumption in the RF-part is very speculative, the feasibility of a much improved baseband, at constant area and power consumption, has been proven in this chapter.

9.6 Conclusions

The theory on conditioning ΣΔ ADCs, as presented in chapter 6, is illustrated with design examples for a Bluetooth receiver. Conclusions are listed below.

The feed forward reference ADC of section 9.2, with signal-conditioning in the decimation filter, achieves a state-of-the-art FOM_{SINAD} of $1 \times 10^{-6} J$ and enables a Bluetooth-compliant receiver without the need for analog signal conditioning.

FOM_{SINAD} is not suited for benchmarking the ADCs of sections 9.3 and 9.4. Instead, FOM_{DR} is used to evaluate the input-referred *DR*: in this respect, the conditioning ADCs surpass state-of-the-art by two orders of magnitude. This is the virtue of integrating signal conditioning into the $\Sigma\Delta$ ADC. Moreover, these ADCs allow a power saving in the consecutive blocks of the channel by a factor of about 20.

The feed forward reference ADC with signal-conditioning in the decimation filter, is recommended for receivers that benefit from a large *SNR* at the output. For receivers that need only a moderate *SNR* (e.g. most wireless communication systems), a conditioning $\Sigma\Delta$ ADC with a filtering STF is recommended.

The FFB-ADC -as compared to the feedback ADC- is capable of achieving a better power/performance balance and features more degrees-of-freedom to optimize the interferer immunity.

Chapter 10

General conclusions

Signal conditioning channels are being digitized in a quest for flexibility and to benefit from technology scaling. While in the signal processing arena digitization has lead to a power saving per computation, this benefit is not self-evident for the conditioning channel. In this context, this book aims at improving the power/performance relation of the conditioning channel by balancing analog and digital signal conditioning.

Digitizing the conditioning channel puts a burden on the data converters. In receiver applications, the bandwidth and linearity requirements on, especially, the digitized A/D channel become very challenging due to the presence of interferers. It is shown that a single-bit, continuous-time ΣΔ modulator with a feed forward loop filter promises the best power/performance balance for the ADC. This topology is used as a reference throughout the book.

Analyzing the power/performance relation of analog circuits, ΣΔ ADCs and digital functions, indicates that a ΣΔ ADC may constitute a power-effective replacement for a cascade of analog blocks. Moreover, it is clear that the power consumption of the ADC becomes dominant in many conditioning channels, even when these are being digitized. In view of the slow advances in the power/performance balance of ADCs -and analog circuits in general- it is concluded that architectural innovation is needed to enable power-effective digitization.

This conclusion is also supported by the analysis of a full-digital conditioning channel: next to the performance challenge for the ADC and the analog reference circuits, even the sample rate in the digital blocks becomes cumbersome. At present, straightforward digitization -by shifting the ADC towards the antenna- is only feasible for narrow-band systems. Even then, the linearity requirement remains very demanding.

As an alternative, the concept of "conditioning ΣΔ ADCs" is proposed: instead of replacing analog conditioning circuits by digital functions, the signal-conditioning can be integrated into the "broader ΣΔ ADC". The key pillar, enabling the conditioning ΣΔ ADC, is the fact that continuous-time ΣΔ ADCs are -to a large extent- immune to interferers.

This characteristic is first exploited in a conditioning ΣΔ ADC with the signal-conditioning in the decimation filter. The available decimation filtering and word-length scaling at the same time provides channel filtering and digital VGA. As compared to the straightforward digitization, this solution reduces the bandwidth and sample rate related problems.

When using a ΣΔ topology that features a filtering signal transfer function, it can be combined with variable gain control of the input signal. The signal-conditioning that is, as such, integrated into the ΣΔ loop, enables a large input-referred *DR* for the ADC while, at the output, the *SINAD* remains moderate. Consequently, this solution provides an additional advantage by relaxing the accuracy required of a.o. the DAC and its references. In topologies with nested feedback, the achievable filtering of the signal transfer function is restricted. In addition, the power/performance balance is less attractive than that of the reference ADC. Therefore, a "filtering-feedback ΣΔ ADC" is proposed as a better solution. It allows for unrestricted design of the signal transfer function and, moreover, it is based on the same low-power topology as the reference ADC.

Appendix A

Overview of published ΣΔ ADCs

Table A.1 lists ΣΔ ADCs published between 2000 and 2004. It includes publications at conferences such as ISSCC, CICC, ESSCIRC and the VLSI Circuits Symposium and papers from the Journal of Solid-State Circuits. Notice:

- all ADCs are implemented in CMOS technology unless "BiC."is mentioned to indicate a BiCMOS implementation. In the second column the feature size of the technology is listed;
- the FOM in the 6th and 7th column are that of eq. 4.4 and eq. 4.6, respectively. The differences between and the use of these FOMs is discussed in section 4.2. It is repeated here that the lower limit on FOM_{SINAD} equals 1.6×10^{-20}J (assuming only noise);
- m, L and N stand for the over-sample ratio, the order of the loop filter and the number of bits in the quantizer, respectively. In case, several numbers are mentioned for L and N, this indicates that a cascaded topology-i.e. a MASH ΣΔ ADC- is reported;
- "CT" and "DT" stand for continuous-time and discrete-time, respectively. "ff" and "fb" indicate whether a loop filter with feed forward or feedback stability compensation is used. If "feedin" is added, branches from the input of the ADC to internal nodes of the loop filter are present, such as in the hybrid topology of section 6.3.4.

Table A.1: *Survey of ΣΔ ADCs published from 2000 until 2004.*

Ref.	Techn μm	BW [kHz]	SINAD [dB]	P [mW]	FOM_{SINAD} $\times 10^{-16}$J	FOM_{ENOB} $\times 10^{-12}$J	m	Type	L	N
[129]	0.5	11	62	1.7	980	75	64	DT, fb+ff	4	1
[130]	0.18	16	70	0.2	13	2.4	48	DT, fb	3	1
[131]	0.35	20	105	55	0.87	9.5	128	DT, ff	5	4
[132]	0.35	20	78	5.6	44	22	256	DT, fb	2	1
[133]	0.5	20	100.3	66	3.1	19	76	DT, fb	5	5
[42]	0.09	20	81	0.13	0.52	0.35	100	DT, fb	3	1
[134]	0.5	24	98	90	5.9	29	128	DT, fb+feedin	2	4
[36]	0.5	25	70	0.135	5.4	1.0	48	CT, fb	3	1
[50]	0.13	25	71	0.25	7.9	1.7	48	CT, fb	3	1
[135]	0.35	25	85	0.95	1.2	1.3	100	DT, fb	3	1
[132]	0.35	50	74	5.6	45	14	102	DT, fb	2	1
[123]	0.13	100	74.1	1.28	5	1.5	500	DT	2	1
[136]	0.13	100	81	2.4	1.9	1.3	195	DT, fb	2	1
[137]	0.18	100	79	5	6.3	3.4	130	CT, fb	3	1
[38]	0.35	100	82	1.8	1.1	0.87	65	CT, ff	4	1
[138]	0.5	135	66	4	74	9.1	48	DT, fb	2	1
[139]	0.4BiC.	180	82	5	1.8	1.3	36	DT, fb	2-2	1-1
[31]	0.18	200	64	1.75	35	3.4	50	CT, fb	2	4
[140]	0.35	200	42.3	12	35000	280	107	DT, fb, BP	2	1
[141]	0.35	200	61	76	3020	210	107	DT		1
[142]	0.18	200	81	30	12	8.2	58	DT, fb	2	6

Ref.	Techn μm	BW [kHz]	SINAD [dB]	P [mW]	FOM_{SINAD} $\times 10^{-16}$J	FOM_{ENOB} $\times 10^{-12}$J	m	Type	L	N
[143]	0.25	200	72	11.5	36	8.8	192	DT, ff	3	1
[2]	0.25	200	82	8	2.5	1.9	52	CT	5	1
[144]	0.8	250	86	30	3	3.7	96	DT, fb	5	1
[102]	0.25	250	77.4	77	56	25	20	DT, fb, BP	2-2	3-3
[145]	0.6	250	94	210	3.3	10	64	DT, ff	5	1
[146]	0.25	270	78.4	9.2	4.9	2.5	48	CT, ff	4	1
[147]	0.35	270	78	24	14	6.8	148	DT, fb, BP	2-2	1-1
[148]	0.35	270	80	56	21	13	148	DT, fb, BP	4	1
[149]	0.35BiC.	270	81	50	15	10	48	DT+CT, fb, BP	6	3
[84]	0.18	271	90	8.1	0.3	0.58	48	CT, ff	5	1
[150]	0.18	276	78	15	8.6	4.2	96	DT, fb	2	3
[125]	0.35BiC.	500	77	12	4.8	2.1	32	DT, fb	2	4
[142]	0.18	625	77	30	10	4.1	18	DT, fb	2	6
[33]	0.18	1000	64	2.2	8.8	0.85	24	CT, fb	2	4
[123]	0.13	1000	58	1.28	20	0.99	50	DT	2	1
[32]	0.18	1000	75.5	4.4	1.2	0.44	32	CT, ff	5	1
[31]	0.18	1000	51	1.75	140	3.0	10	CT, fb, BP	2	4
[39]	0.65	1000	56.7	21.8	470	20	50	CT, fb+feedin, BP	2	1
[151]	0.18	1000	59	2	25	1.4	32	CT, ff	4	1
[28]	0.18	1000	77.3	6	1.1	0.5	141	CT, ff	3	3.2
[152]	0.18	1000	88	230	3.6	5.6	29	DT, fb	2-1-1	1-1-2
[124]	0.5	1100	84	62	2.2	2.2	16	CT, ff+feedin	3	5

Ref.	Techn μm	BW [kHz]	SINAD [dB]	P [mW]	FOM_{SINAD} $\times 10^{-16}$J	FOM_{ENOB} $\times 10^{-12}$J	m	Type	L	N
[43]	0.13	1100	78	7	1	0.49	192	DT, fb+feedin	2	3
[84]	0.18	1230	83	8.7	0.35	0.31	31	CT, ff	5	1
[147]	0.35	1250	75	37	9.4	3.2	32	DT, fb, BP	4-4	1-1
[153]	0.25	1250	80	100	8	4.9	2	DT, ff	4	4
[154]	0.5	1250	90	105	0.84	1.6	8	DT, fb	2-1-1	4-4-4
[49]	0.65	1250	89	295	3	5.1	24	DT, fb	2-1-1	1-1-4
[147]	0.35	1762	69	37	26	4.6	22	DT, fb, BP	4-4	1-1
[142]	0.18	1920	70	50	26	5.0	6	DT, fb	2	6
[155]	0.13	1920	50.9	1.5	64	1.4	16	CT, fb	2	2.3
[136]	0.13	1920	64	4.3	8.9	0.86	10	DT, fb	2-1	1-2.3
[156]	0.12	2000	60	3	15	0.92	26	CT, ff	3	3
[123]	0.13	2000	45.2	1.28	190	2.2	25	DT	2	1
[153]	0.25	2000	74	105	21	6.4	12	DT, ff	4	4
[157]	0.18	2000	82	150	4.7	3.6	8	DT, ff+fb, BP	5	4
[37]	0.18	2000	68	3.3	2.6	0.4	38	CT, ff	4	2
[158]	0.5	2000	87	150	1.5	2.0	16	DT,fb	2-2-1	5-3-3
[159]	0.25	2000	70	110	55	11	16	DT	2-2	1-1
[160]	0.18	2500	69	150	76	13	24	DT, BP	2-2-2	3-5-5
[161]	0.6	3100	56	16	130	5.0	64	CT, ff+fb	5	1
[147]	0.35	3840	48	38	1570	24	10	DT, fb, BP	4-4	1-1
[84]	0.18	3840	72	9.5	1.6	0.38	20	CT, ff	5	1
[148]	0.35	3840	42	56	9200	71	10	DT, fb, BP	4	1

Ref.	Techn μm	BW [kHz]	SINAD [dB]	P [mW]	FOM_{SINAD} $\times 10^{-16}$J	FOM_{ENOB} $\times 10^{-12}$J	m	Type	L	N
[143]	0.25	3840	52	13.5	220	5.4	24	DT, ff	3	1
[162]	0.65	6250	67	380	120	17	8	DT, fb	3	4
[163]	0.18	12000	72	200	11	2.6	8	DT, fb	5	4
[34]	0.13	15000	63.7	70	20	1.9	10	CT, ff+fb	4	4
[85]	0.18	20000	56	122	150	5.9	8	CT, ff	2-2	4-4
[6]	0.13	20000	50	80	400	7.7	4	DT, fb	2-2	1.5-4
[6]	0.13	40000	50	106	270	5.1	4	DT, fb	2-2	1.5-4

Appendix B

Power/performance relation of analog circuits

First order relations between the current consumption on one side and noise, distortion and signal level on the other side are derived for the topologies with differential pair in figure 4.2. It is assumed that the transconductance of the input transistors constitute the dominant noise and distortion source.

B.1 Simple differential pair

Some well-known equations for noise and distortion of the simple differential pair of figure 4.2.a can be calculated in a straight-forward way.

$$DR = \frac{g_m \hat{v}_{\mathrm{IN}}^2/2}{(8kT\gamma\, BW)} \tag{B.1}$$

$$IM_3 = \frac{3}{32}\left(\frac{\hat{v}_{\mathrm{IN}}}{v_{\mathrm{GT}}}\right)^2 \tag{B.2}$$

Notice $\hat{v}_{\mathrm{IN}}$ refers to the amplitude of the input signal. The parameter γ is the noise-excess factor. In present CMOS technologies $4kT\gamma$ equals $2.1 \times 10^{-20} J$.

From the above equations a dependence of the quiescent current on the noise and distortion metrics can be derived. It should be mentioned that often the input signals to the conditioning channel are very small and distortion is not a problem for the first stages. In that case, the overdrive voltage v_{GT} can be chosen close to weak inversion (yielding the highest transconductance for the current spent) instead of the value as calculated in (B.2). The current consumption then is independent of the linearity requirement. In order to generalize the results, a parameter α has been introduced: α becomes zero for relaxed distortion demands, α equals one in case v_{GT} is set by equations B.2:

$$I \sim \frac{DR\, BW}{\hat{v}_{\mathrm{IN}}^2}\left(\frac{\hat{v}_{\mathrm{IN}}}{\sqrt{IM_3}}\right)^{\alpha} \quad (\alpha = 0, 1) \tag{B.3}$$

B.2 Differential pair in a global feed-back configuration

For stages further on in the channel linearity can be a problem and the v_{GT} value required from (B.2) may become too high. Instead, negative feed-back can be applied to linearize the circuit. Moreover, the circuit becomes less sensitive to distortion of the output impedance. The penalty of feed-back is in the higher bandwidth requirement. Referring to figure 4.2.c, it is assumed that the closed loop gain is large such that $R_{in} << R_{fb}$. Within the signal bandwidth the open loop gain $|A(s)|$ on its turn is assumed to be much larger than the closed loop gain R_{fb}/R_{in}, hence: $g_m R_{in} >> 1$. Therefore, R_{in} constitutes the dominant noise impedance:

$$DR = \frac{\hat{v}_{\mathrm{IN}}^2/2}{8kTR_{in}BW} \tag{B.4}$$

Distortion is determined by the error signal $\hat{v}_\epsilon$ at the virtual ground node compared to the v_{GT} of the input transistor. It can be calculated that:

$$\hat{v}_\epsilon = \hat{v}_{\mathrm{IN}} \frac{R_{fb}}{A(s)R_{in}} \tag{B.5}$$

A third-order intermodulation component is generated by the non-linearity of the differential pair. It can be represented by an equivalent signal source of value:

$$\frac{3}{32 v_{\mathrm{GT}}^2} \hat{v}_\epsilon^3 \tag{B.6}$$

at the virtual ground node with [164]. The input-referred equivalent distortion source is larger by a factor $1 + R_{in}/R_{fb}$ which is close to 1. It is compared to the input signal yielding the IM_3 distortion:

$$IM_3 = \frac{3}{32} \left(\frac{\hat{v}_{\mathrm{IN}}}{v_{\mathrm{GT}}} \right)^2 \left(\frac{R_{fb}}{A(s)R_{in}} \right)^3 \tag{B.7}$$

This equation can be simplified in case of an operational transconductance amplifier. Within the bandwidth of interest $A(s)$ equals $g_m R_{fb}$, hence:

$$IM_3 = \frac{3}{32} \left(\frac{\hat{v}_{\mathrm{IN}}}{v_{\mathrm{GT}}} \right)^2 \left(\frac{1}{g_m R_{in}} \right)^3 \tag{B.8}$$

If the input transistors are biased near weak-inversion :

- the linear input range of the differential pair decreases
- the intermodulation component of eq. B.6 reduces because $A(s)$ increases

The latter effect is dominant because of the cubic relation. Hence distortion decreases.

From equation B.4 the maximum input resistance R_{in} can be calculated. This value is filled out in equation B.7. As the open loop gain $A(s)$ is proportional to the quiescent current, the following power/performance relation can be derived:

$$I \sim \frac{DR\,BW}{\hat{v}_{\mathrm{IN}}^2} \left(\frac{\hat{v}_{\mathrm{IN}}}{\sqrt{IM_3}} \right)^\alpha \quad (\alpha = 2/3) \tag{B.9}$$

The α-parameter has been used in view of comparison with the results for the simple differential pair.

B.3 Degenerated differential pair

An alternative solution in case of challenging linearity requirements is in applying degeneration to the differential pair (figure 4.2.c). In fact, degeneration is a case of local feed-back and the same power/performance relations as expressed by equation (B.9) can be calculated.

Appendix C

Power/performance relation of digital filters

In section 4.5.2 the power consumption of a digital filter is expressed as a function of a.o. the sample rate and the capacitance. Here, a relation between the capacitance and the filter specifications is derived. This is done per clock domain, i.e. per decimation stage. Next, the filter specifications are written in terms of the parameters of the conditioning channel, i.e. *ENOB*, mf_s and f_s. Combining these relations, an upper limit on the power/performance relation of decimation filters results. The analysis is conducted assuming the direct-form FIR implementation of fig. C.1 for all decimation sections. Obviously, this implementation cannot be used for mapping the CIC filters. This is commented afterwards.

C.1 Analysis of the filter topology

The implementation of figure C.1 consists of a shift register operating at the high sample rate and a computational part operating at the decimated frequency. It is assumed that the filter coefficients are in a Canonic Signed Digit (CSD) format. In that case, the multiplication can be rewritten in terms of additions only[1]. Furthermore, the following nomenclature is used for the filter parameters:

N_i	= number of taps of decimation stage i
W_i	= word-length at the input of stage i
$m_i f_s$	= sample rate at the input of stage i (m_1 equals m)
p_i	= average number of additions per tap of stage i (assuming every computation can be written as an addition)

[1] In the CSD format a number is expressed as a sum of powers of 2. This allows nested multiplication: partial products are calculated using simple bit shifts and are summed afterwards. Hence, the complexity of a multiplication is then reduced to that of adding a number of partial products.

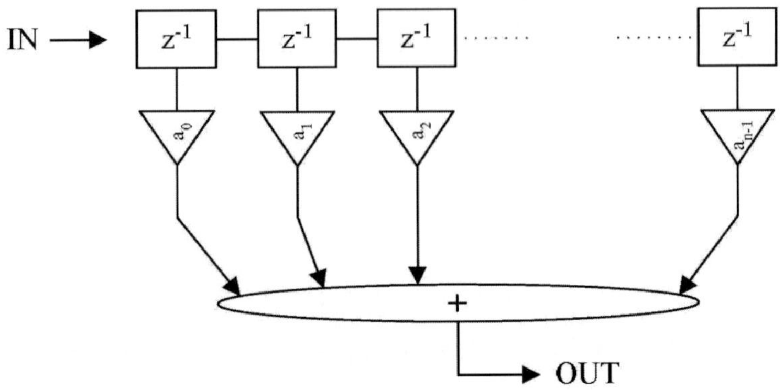

Figure C.1: *Direct-form FIR implementation*

The current consumption in the shift register of stage i can then be expressed as:

$$I_{i,shift} \sim m_i f_s N_i W_i \tag{C.1}$$

The current consumption of the computational part obeys:

$$I_{i,comp} \sim m_{i-1} f_s N_i W_i p_i \tag{C.2}$$

C.2 Calculation of filter parameters

The filter parameters N_i, p_i and W_i are expressed in terms of the parameters of the conditioning channel.

N_i: The number of taps N_i can be estimated using the empirical formula by Kaiser:

$$N_i = \frac{-10\log\left(\delta_p \delta_s\right) - 13}{\frac{14.6\Delta f_i}{m_i f_s}} \tag{C.3}$$

Δf_i is the transition band of the ith decimation stage. The pass-band ripple δ_p was assumed constant and it approximates 1 anyhow. Hence, the Kaiser formula indicates that the number of taps is proportional to the required stop-band suppression (expressed in dB) divided by the relative transition band of the filter stage. The required stop-band suppression (expressed in dB) is proportional to the $ENOB$[2], hence:

$$N_i \sim \frac{ENOB}{\frac{\Delta f_i}{m_i f_s}} \tag{C.4}$$

[2] As the resolution requirements increase the quantization noise must be suppressed deeper.

p_i: The number of partial additions p_i (per filter tap) depends on the accuracy of the coefficient. As this accuracy decreases the frequency response of the filter degrades and more quantization noise folds back. A rule of thumb for the number of CSD digits per coefficient -and thus for the number of additions of partial products- is given in reference [165]: one CSD per 20dB of stop-band suppression should be counted, hence:

$$p_i \sim ENOB \tag{C.5}$$

W_i: The word-length W_i equals 1 in the input stage (considering only single-bit $\Sigma\Delta$ conversion). In the last stage it approximates *ENOB*. In the intermediate stages it has an in-between value.

C.3 Calculation of power consumption

Relations C.1 and C.2 can now be filled out for the various decimation stages.

First decimation stage: The contributions of the shift register and the computational part are added. N_1 is calculated from eq. C.4 with a relative transition band of almost 1/2. W_i equals 1 for single-bit $\Sigma\Delta$ A/D conversion and p_1 follows from eq. C.5.

$$I_1 \sim ENOB\ mf_s + ENOB^2\ mf_s \leq O\left(ENOB^2\ mf_s\right) \tag{C.6}$$

for large *ENOB*.

Last decimation stage: The relative transition band is set by the application (as discussed on page 48). The word-length of the data about equals the required *ENOB* and the number of additions is again determined by the stop-band suppression.

$$I_{last} \sim \frac{ENOB^2}{\frac{\Delta f}{2f_s}} 2f_s + \frac{ENOB^3}{\frac{\Delta f}{2f_s}} f_s \leq O\left(\frac{ENOB^3}{\frac{\Delta f}{2f_s}} f_s\right) \tag{C.7}$$

Notice *ENOB* appears in a cubic relation: the number of filter taps, the word-length of the coefficients and of the data are all proportional to *ENOB*. Hence, the last decimation stage is normally the largest in terms of die area. Also, it must achieve the smallest relative transition band of all sections. This strongly affects the number of filter taps, the area and the power consumption. Although, this stage is sampled at an m times lower frequency than the first stage its power consumption can be as -or even more- important due to its complexity.

Intermediate stages: As the sample rate gets decimated the complexity (i.e. number of taps, data word-length and coefficient word-length) of the filter stage increases. Hence, the current consumption of the intermediate stages is expected to follow a relation in between that of the first (eq. C.6) and the last (eq. C.7) stage.

Overall current consumption: The contributions of the various stages should be summed in order to calculate the overall current consumption. The number of decimation stages is a logarithmic function of the over-sample ratio m. The base of the logarithm depends on the decimation factors used; e.g. in case all stages decimate by a factor of 2 about $\ln_2 m$ stages are needed.

In case the last decimation stage dominates the power consumption, a cubic dependence on ENOB is valid while the over-sample ratio is of less importance. In case the first decimation stage dominates the input frequency mf_s is of utmost importance while the dependence on ENOB is only quadratic. An upper-limit on the dependence of the overall current consumption on the parameters of the conditioning channel is expressed by the following relation:

$$I_{worst-case} \sim O\left(\frac{ENOB^3}{\frac{\Delta f}{2f_s}} mf_s \ln m\right) \tag{C.8}$$

Of course, this limit combines a worst-case combination of dependencies and is pessimistic. Also, it only gives relative results. Hence, it is only intended for extrapolation of the power consumption of a specific architecture to modified performance specifications.

C.4 Extension to other implementations

The power/performance analysis has been conducted for the direct-form FIR implementation of figure C.1. It can easily be understood that transposed-form, poly-phase implementations, etc. follow the same relation with a different proportionality parameter. For the implementation of a CIC filter a similar analysis can be conducted. The result is the same as that of eq. C.8 except for the dependence on *ENOB*. Here, *ENOB* only occurs in a quadratic relation because CIC filters do not use coefficients (the data is integrated instead).

Appendix D

Third-order distortion in analog circuits and ΣΔ ADCs

Based on the models in fig. D.1.a and b an expression for the third-order distortion in analog circuits and in a ΣΔ ADC is derived. The model in fig. D.1.a represents an analog circuit with feedback. The model in fig. D.1.b represents a ΣΔ ADC: a bitstream signal is fed back. The output of the quantizer can be scaled such that the linearized DAC gain equals 1. It is shown that the topology of fig. D.1.b results in a two times larger third-order distortion, even if the same input stage (thus the same non-linearity) is used in both cases. Assuming differential operation, the input transconductor performs the following V-to-I conversion:

$$i \approx h_1 v_\epsilon + h_3 v_\epsilon^3 \tag{D.1}$$

where v_ϵ equals the difference between output and input (assuming a gain of 1 in the feedback). h_1 and h_3 are the transfer coefficients for the fundamental and for the 3rd-order distortion component respectively. The higher-order distortion components are neglected.

Furthermore, a sine-wave input is assumed and the following relations are used:

$$v_{IN} = \sin x \tag{D.2}$$

$$v_{IN}^2 = 1/2 - 1/2\ \cos 2x \tag{D.3}$$

$$v_{IN}^3 = 3/4 \sin x - 1/4\ \sin 3x \tag{D.4}$$

In the following, the amplitude of the 3rd-order distortion component of the input stage is calculated for the model of fig. D.1.a and for that of fig. D.1.b. Assuming equal loop gain in both cases, the transfer function from this node to the output node is the same. Hence, it is sufficient to calculate the h_3-coefficient of the input stage and compare the result for model a and b.

Analog circuit (fig. D.1.a):

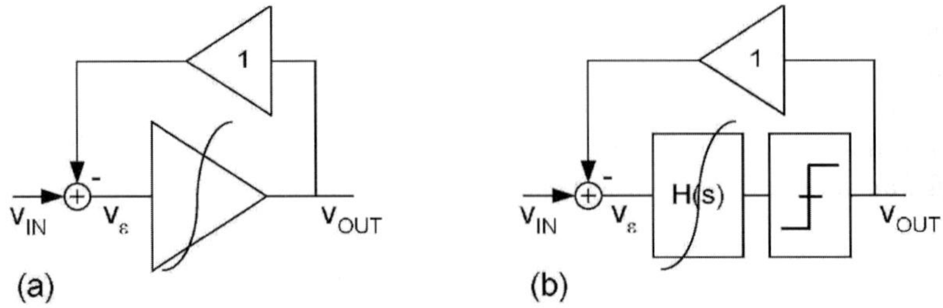

Figure D.1: *Functional model of a non-linear, analog circuit with feedback (a) and of a non-linear ΣΔ ADC (b)*

Assuming:

$$v_{OUT} = (1 + \delta)v_{IN} \tag{D.5}$$

with δ a small number. Using eq. D.4, the last term in eq. D.1 is expanded:

$$h_3 v_\epsilon^3 = \frac{h_3}{4}\delta^3 v_{IN}^3 \tag{D.6}$$

Hence, the amplitude of 3rd-order distortion component of the input stage equals:

$$|\frac{h_3}{4}| \tag{D.7}$$

ΣΔ ADC (fig. D.1.b):
Again, the last term in eq. D.1 is expanded:

$$h_3 v_\epsilon^3 = h_3 v_{IN}^3 - 3h_3 v_{IN}^2 v_{OUT} + 3h_3 v_{IN} v_{OUT}^2 - h_3 v_{OUT}^3 \tag{D.8}$$

Since v_{OUT} is a bitstream encoded signal, the last two terms do not contribute any third-harmonic distortion [40]:

- the square of a bitstream signal yields a constant
- consequently, the 3rd power of a bitstream signal yields the original bitstream

Hence, the 3rd-order distortion component of the input stage is calculated from the first two terms of eq. D.8. The low-frequency content of the output bitstream equals the input sine wave (except for the error δ), i.e. eq. D.5 is valid for low-frequency inputs. Furthermore, relations eq. D.2 upto eq. D.4 are substituted, yielding the amplitude of the 3rd-order distortion component:

$$|\frac{h_3}{4} - \frac{3h_3}{4}(1 + \delta)| \approx |\frac{h_3}{2}| \tag{D.9}$$

if δ is small. Note that this amplitude is twice larger than that of eq. D.7.

In the models of fig. D.1.a and b the same non-linearity is assumed for the input stage of the analog circuit and for the ΣΔ ADC. The loop gain and, therefore, the error δ *on the output signal are assumed to be the same as well. Still, the* 3rd*-order distortion component is a factor two higher for the ΣΔ ADC. This is due to the fact that a bitstream signal is fed back instead of a sine wave.*

Appendix E

Power consumption in a data interface

The power consumption in an analog and a digital inter-die interface is studied. The physical interface between the ICs may consist of bond-pads, bond wires, PCB tracks, etc. It should be modeled by an RLC-network with a certain filtering -or even resonating- transfer. The influence of the inductance is disregarded here. It is assumed that the resonance occurs at frequencies way beyond the band of interest and/or that the resonance is properly damped.

Furthermore, only the power consumed at the transmit side of the link is considered. The power at the receiver side is much smaller because of an easier load (i.e. a smaller capacitance and a higher resistance).

In all cases a differential interface is assumed.

E.1 Analog data interface

A model of the transmit side of the analog link is depicted in figure E.1. Within the bandwidth of this link:

$$\frac{1}{2\pi R_L C_L} \tag{E.1}$$

The peak-to-peak voltage swing V_{pp} equals:

$$V_{pp} = I\, R_L \tag{E.2}$$

It is determined by the quiescent current I through the load resistor R_L. Noise is contributed both by the transconductance of the transmit stage and by the resistive load:

$$\overline{v^2}_{n,eq} = 4kT\ BW\left(\gamma g_m R_L^2 + R_L\right) \tag{E.3}$$

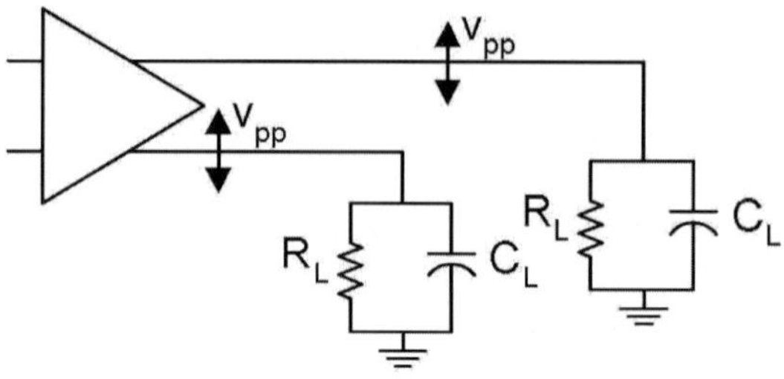

Figure E.1: *Transmit model for an analog interface*

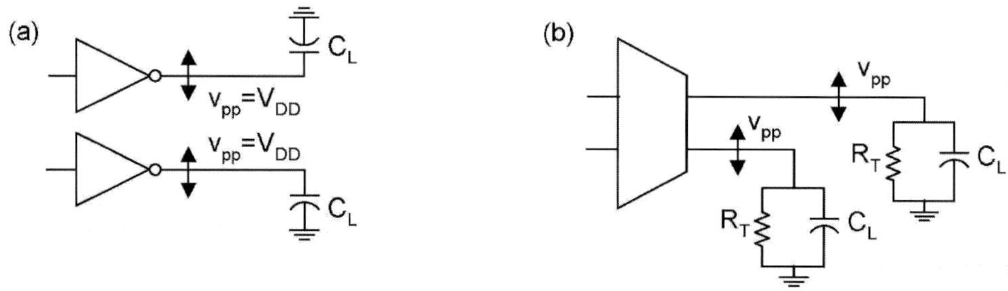

Figure E.2: *Transmit model for full-swing (a) and low-swing (b) digital interface [89]*

Assuming a gain $g_m R_L$ larger than 1 the contribution of the transconductance is dominant, hence:

$$DR = \frac{I \mid v_{GT} \mid}{16kT\gamma BW} \tag{E.4}$$

The dynamic range DR can be expressed as a function of the number of bits:

$$DR = \frac{3}{2}\, 2^{2N} \tag{E.5}$$

From the above equations the current consumption of the analog transmitter can be calculated. Normally, a substantial margin is foreseen on this value to account for distortion and interference injected in the link. Hence, the result is given in terms of a proportionality relation:

$$I \sim 2^{2N} BW \tag{E.6}$$

E.2 Digital data interface

This analysis is partly taken from [89] and is recapitulated and supplemented here. A full-swing and a low-swing (or slew-rate controlled) interface are distinguished (figure E.2.a and b respectively). A low-swing digital interface generates less interference.

It should be noticed that a digital communication link requires synchronization at the receiver side. This can be done by either transmitting the clock or recovering it from the data. The associated power is not considered here.

Furthermore, the average switching frequency f_{sw} is introduced: it is defined as the inverse of the average number of *0−1* transitions in the bit stream. Since f_{sw} is data-dependent it results that the power consumption in a digital link is data dependent as well.

Full-swing interface A full-swing interface normally uses an inverter type of transmitter as depicted in figure E.2.a. The bandwidth of the link is limited by the slew-rate of the inverters for the specified load. It can be calculated that the average current consumption of the pseudo-differential transmitter equals:

$$I_{full-swing} = C_L Vdd\, N\, f_{sw} \tag{E.7}$$

The parameter N indicates the number of bits that must be transmitted either via some parallel links or via the same link after serializing the data.

Low-swing interface For the low-swing interface a termination resistor R_T is introduced. Its value is either standardized or it follows from the following bandwidth requirements:

$$\frac{1}{2\pi R_T C_L} \gg f_{sw} \qquad \text{for a parallel interface} \tag{E.8}$$

$$\frac{1}{2\pi R_T C_L} \gg N f_{sw} \qquad \text{for a serialized interface} \tag{E.9}$$

Part of the current flows into R_T, the other part is charging and discharging the load capacitor C_L. The current consumption in the interface is given by:

$$I_{low-swing} = (R_T^{-1} + \frac{C_L\, f_{sw}}{2}) V_{pp} N \qquad \text{for a parallel interface} \tag{E.10}$$

$$I_{low-swing} = (R_T^{-1} + \frac{C_L\, N f_{sw}}{2}) V_{pp} \qquad \text{for a serialized interface} \tag{E.11}$$

In both cases, the capacitive part of the current consumption (i.e. the dynamic current) can be neglected compared to the dissipation in the termination resistor (i.e. the static current) because of the bandwidth requirement (eq. E.9):

$$I_{low-swing} \cong R_T^{-1} V_{pp} N \qquad \text{for a parallel interface} \tag{E.12}$$

$$I_{low-swing} \cong R_T^{-1} V_{pp} \qquad \text{for a serialized interface} \tag{E.13}$$

Serializing the parallel bitstreams is favorable for power consumption since the available bandwidth is used more effectively. Comparing the result to that for the full-swing interface (eq. E.7) it can be concluded that, in case of a low-resolution link, the full-swing interface is lower power because it does not use any static current. For higher data rates or a large load capacitance a low-swing interface should be favored.

In case the value of R_T is fixed by a standard or application the current consumption in the low-swing link is independent of f_{sw} as long as the bandwidth requirements is fulfilled. It scales with the number of parallel bits.

In case R_T is not standardized, the link can be designed for a minimum current consumption by choosing R_T as high as possible. The upper limit on R_T is given by eq. E.9. When scaling R_T to this limit, the dynamic and the static contribution in eq. E.13 become equally important. Eq. E.11 can be simplified such that only the dependence on the number of bits N and on the average switching frequency f_{sw} remains:

$$I_{low-swing} \sim N f_{sw} \tag{E.14}$$

This relation holds both for the parallel and the serialized link. Also, the scaling with N and f_{sw} is the same as for the full-swing interface (eq. E.7) with a capacitive load only.

References

[1] M. Pelgrom and M. Vertregt, "CMOS technology for mixed-signal IC's," *Solid-State Electron.*, vol. 41, no. 7, pp. 967–974, 1997.

[2] E. van der Zwan, K. Philips, and C. Bastiaansen, "A 10.7MHz IF-to-baseband ΣΔ A/D conversion system for AM/FM radio receivers," *IEEE J. Solid-State Circuits*, vol. 35, no. 12, pp. 1810–1819, Dec. 2000.

[3] W. Sansen, M. Steyaert, V. Peluso, and E. Peeters, "Toward sub 1V analog Integrated Circuits in submicron standard CMOS technologies," in *ISSCC Digest of Tech. Papers*, vol. 41, Feb. 1998, pp. 186–187, 435.

[4] A. Abidi, G. Pottie, and W. Kaiser, "Power-conscious design of wireless circuits and systems," in *Proc. IEEE*, vol. 88, Oct. 2000, pp. 1528–1545.

[5] A. Annema, "Analog circuit performance and process scaling," *IEEE Trans. Circuits and Systems II*, vol. 46, no. 6, pp. 711–715, June 1999.

[6] A. Tabatabaei, K. Onodera, M. Zargari, H. Samavati, and D. Su, "A dual channel ΣΔ ADC with 40MHz aggregate signal bandwidth," in *ISSCC Digest of Tech. Papers*, vol. 46, Feb. 2003, pp. 66–67, 478.

[7] E. van der Zwan and E. C. Dijkmans, "A 0.2mW CMOS ΣΔ modulator for speech coding with 80dB dynamic range," *IEEE J. Solid-State Circuits*, vol. 31, no. 12, pp. 1873–1880, Dec. 1996.

[8] A. Annema, "CMOS technology evolution and analog circuit performance," in *Proc. 12th ProRISC Workshop*, 2001, pp. 245–249.

[9] A.J.Scholten, L. Tiemeijer, R. van Langevelde, R. Havens, A. T. A. Z. van Duijnhoven, and V. Venezia, "Noise modeling for RF CMOS circuit simulation," *IEEE Trans. Electron. Devices*, vol. 50, no. 3, pp. 618–632, Mar. 2003.

[10] R. van Langevelde, "A compact MOSFET model for distortion analysis in analog circuit design," Ph.D. dissertation, Eindhoven Univ. of Techn., The Netherlands, Nov. 1998.

[11] A. Scholten, H. Tromp, L. Tiemeijer, R. van Langevelde, R. Havens, P. de Vreede, R. Roes, P. Woerlee, A. Montree, and D. Klaassen, "Accurate thermal noise model for deep-submicron CMOS," in *IEDM Digest of Tech. Papers*, Dec. 1999, pp. 155–158.

[12] W. Sansen, "Analog circuit design in scaled CMOS technology," in *VLSI Circuits Digest of Tech. Papers*, June 1996, pp. 8–11.

[13] Q. Huang, F. Piazza, P. Orsatti, and T. Ohguro, "The impact of scaling down to deep submicron on CMOS RF circuits," *IEEE J. Solid-State Circuits*, vol. 33, no. 7, pp. 1023–1036, July 1998.

[14] International Technology Roadmap for Semiconductors 2003 edition. ITRS. [Online]. Available: http://public.itrs.net/Files/2003ITRS/Home2003.htm

[15] *Technical specification group radio access networks; UE radio transmission and reception (FDD)*, 3rd generation partnership project Std. release 99 3G TS 25.213, v3.2.0.

[16] B. Minnis and P. Moore, "A highly digitized multimode receiver architecture for 3G mobiles," *IEEE Trans. Vehicular Technology*, vol. 52, no. 3, pp. 637–653, May 2002.

[17] ——, *A qualitative review of UMTS transmitter architectures*. PRL Tech. Note 4061, Philips internal report, 2001.

[18] A. Caswell and I. Ozerin, *Dimensioning a polar loop for a UMTS transmitter*. PRL Tech. Note 4099, Philips internal report, 2001.

[19] K. Philips and B. Minnis, *Considerations on an A-to-D converter for UMTS*. Nat.Lab. Tech. Note 425, Philips internal report, 1999.

[20] B. Murmann and B. Boser, "Digitally assisted analog integrated circuits," *DSP*, vol. 2, no. 1, pp. 2040–2050, Mar. 2004.

[21] "French patent presentation," 987283, May 22, 1948.

[22] F. de Jager, "Delta modulation, a method of PCM transmission using the one unit code," *Philips Res. Rep.*, vol. 7, pp. 442–466, 1952.

[23] C. Cutler, "Transmission systems employing quantization," U.S. Patent2,927,962, March 8, 1960.

[24] H. Inose, Y. Yasuda, and J. Murakami, "A telemetering system by code modulation-$\Delta\Sigma$ modulation," *IRE Trans. Space Electron. Telemetry*, vol. SET-8, pp. 204–2–9, Sept. 1962.

[25] S. Norsworthy, R. Schreier, and G. Temes, *Delta-Sigma Data Converters*. New York: IEEE Press, 1996.

[26] J. van Engelen and R. van de Plassche, *Bandpass Sigma delta modulators.* Kluwer Academic Publishers, 1999.

[27] J. Cherry and W. Snelgrove, *Continuous-time Delta-Sigma Modulators for high-speed A/D conversion.* Norwell, MA: Kluwer Academic Publishers, 2000.

[28] B. Putter, "ΣΔ ADC with finite impulse response feedback DAC," in *ISSCC Digest of Tech. Papers*, Feb. 2004, pp. 76–77.

[29] J. Yu and F. Maloberti, "A low-power multi-bit ΔΣ modulator in 90nm digital CMOS without DEM," in *ISSCC Digest of Tech. Papers*, Feb. 2005, pp. 168–169, 591.

[30] R. van Veldhoven, "A tri-mode continuous-time ΣΔ modulator with switched-capacitor feedback DAC for a GSM-EDGE/CDMA2000/UMTS receiver," in *ISSCC Digest of Tech. Papers*, Feb. 2002, pp. 60–61, 477.

[31] M. Kappes, H. Jensen, and T. Gloerstad, "A versatile 1.75mW CMOS continuous-time Δ-Σ ADC with 75dB dynamic range for wireless applications," in *ESSCIRC Digest of Tech. Papers*, Sept. 2002, pp. 279–282.

[32] K. Philips, "A 4.4mW 76dB complex ΣΔ ADC for Bluetooth receivers," in *ISSCC Digest of Tech. Papers*, Feb. 2003, pp. 64–65,478.

[33] M. Kappes, "A 2.2-mW CMOS bandpass continuous-time multibit Δ-Σ ADC with 68dB of dynamic range and 1MHz bandwidth for wireless applications," *IEEE J. Solid-State Circuits*, vol. 38, no. 7, pp. 1098–1104, July 2003.

[34] S. Paton, A. D. Giandomenico, L. Hernandez, A. Wiesbauer, T. Potscher, and M. Clara, "A 70mW 300MHz CMOS continuous-time ΣΔ ADC with 15MHz bandwidth and 11 bits of resolution," *IEEE J. Solid-State Circuits*, vol. 39, no. 7, pp. 1056–1063, July 2004.

[35] P. Vancorenland, P. Coppejans, W. D. Cock, and M. Steyaert, "A quadrature digital downconverter," in *CICC Digest of Tech. Papers*, 2002, pp. 235–238.

[36] F. Gerfers, M. Ortmanns, and Y. Manoli, "A 1.5V 12-bit power-efficient continuous-time third-order ΣΔ modulator," *IEEE J. Solid-State Circuits*, vol. 38, no. 8, pp. 1343–1352, Aug. 2003.

[37] R. van Veldhoven, B. Minnis, and H. Hegt, "A 3.3-mW ΣΔ modulator for UMTS in 0.18μm CMOS with 70-dB dynamic range in 2-MHz bandwidth," *IEEE J. Solid-State Circuits*, vol. 37, no. 12, pp. 1645–1661, Dec. 2002.

[38] L. Breems, E. van der Zwan, and J. Huijsing, "A 1.8-mW CMOS ΣΔ modulator with integrated mixer for A/D conversion of IF signals," *IEEE J. Solid-State Circuits*, vol. 35, no. 4, pp. 468 –475, Apr. 2000.

[39] F. Henkel, U. Langmann, A. Hanke, S. Heinen, and E. Wagner, "A 1-MHz-bandwidth second-order continuous-time quadrature bandpass sigma-delta modulator for low-IF radio receivers," *IEEE J. Solid-State Circuits*, vol. 37, no. 12, pp. 1628–1635, Dec. 2002.

[40] L. Breems and J. Huijsing, *Continuous-Time Sigma-Delta Modulation for A/D Conversion in Radio Receivers.* Kluwer Academic Publishers, 2001.

[41] H. van der Ploeg, "Calibration techniques in two-step A/D converters," Ph.D. dissertation, Univ. Twente, The Netherlands, Mar. 2005.

[42] L. Yao, M. Steyaert, and W. Sansen, "A 1V 88dB 20kHz ΣΔ modulator in 90nm CMOS," in *ISSCC Digest of Tech. Papers*, Feb. 2004, pp. 80–81, 514.

[43] R. Gaggl, M. Inversi, and A. Wiesbauer, "A power optimized 14-bit SC ΔΣ modulator for ADSL CO applications," in *ISSCC Digest of Tech. Papers*, Feb. 2004, pp. 82–83, 514.

[44] W. Gao, O. Shoaei, and W. Snelgrove, "Excess loop delay effects in continuous-time delta-sigma modulators and the compensation solution," in *Proc. Int. Symp. Circuits Syst.*, vol. 1, 1997, pp. 65–68.

[45] J. Cherry and W. Snelgrove, "Excess loop delay in continuous-time delta-sigma modulators," *IEEE Trans. Circuits and Systems II*, vol. 46, no. 4, pp. 376–389, Apr. 1999.

[46] G. Hurkx, P. Agarwal, R. Dekker, E. van der Heijden, and H. Veenstra, "RF figures-of-merit for process optimization," *IEEE Trans. Electron. Devices*, vol. 51, no. 12, pp. 2121 – 2128, Dec. 2004.

[47] T. Burger and Q. Huang, "A 13.5mW, 185MSample/s ΔΣ -modulator for UMTS/GSM dual-standard IF reception," in *ISSCC Digest of Tech. Papers*, Feb. 2002, pp. 44–45, 427.

[48] H. Tao and J. M. Khoury, "A 100MHz IF, 400MSample/s CMOS direct-conversion bandpass ΣΔ modulator," in *ISSCC Digest of Tech. Papers*, Feb. 1999, pp. 60–61, 445.

[49] Y. Geerts, M. Steyaert, and W. Sansen, "A 2.5MSample/s multi-bit ΔΣ CMOS ADC with 95dB SNR," *IEEE J. Solid-State Circuits*, vol. 35, no. 12, pp. 1829–1839, Dec. 2000.

[50] M. Ortmanns and F. Gerfers, "A continuous-time ΣΔ modulator with reduced jitter sensitivity," in *ESSCIRC Digest of Tech. Papers*, Sept. 2002, pp. 287–290.

[51] S. Luschas and H. Lee, "High-speed ΣΔ modulators with reduced timing jitter sensitivity," *IEEE Trans. Circuits and Systems II*, vol. 49, no. 11, pp. 712–720, 2002.

[52] R. Adams and K. Nguyen, "A 113dB SNR oversampling DAC with segmented noise-shaping scrambling," in *ISSCC Digest of Tech. Papers*, Feb. 1998, pp. 1871 – 1878.

[53] P. Allen, "Slew induced distortion in operational amplifiers," *IEEE J. Solid-State Circuits*, vol. 12, no. 1, pp. 39–44, Feb. 1977.

[54] W. Sansen, H. Qiuting, and K. Halonen, "Transient analysis of charge transfer in SC filters - gain error and distortion," *IEEE J. Solid-State Circuits*, vol. 22, no. 2, pp. 268–276, Apr. 1987.

[55] E. Vittoz, "Future of analog in the VLSI environment," *IEEE Int. Symp. on Circuits and Systems*, vol. 2, pp. 1372 –1375, May 1990.

[56] J. Meindl and J. Davis, "*The* fundamental limit on binary switching energy for terascale integration (TSI)," *IEEE J. Solid-State Circuits*, vol. 35, no. 10, pp. 1515–1516, Oct. 2000.

[57] E. Dijkstra, O. Nys, and E. Blumenkrantz, *Low power Oversampled A/D converters.* Kluwer Academic Publishers, 1997, pp. 89–103.

[58] A.-J. Annema, B. Nauta, R. van Langevelde, and H. Tuinhout, "Analog circuits in ultra-deep-submicron CMOS," *IEEE J. Solid-State Circuits*, vol. 40, no. 1, pp. 132 – 143, Jan. 2005.

[59] M. Steyaert, V. Peluso, J. Bastos, P. Kinget, and W.Sansen, "Custom analog low power design: the problem of low voltage and mismatch," in *CICC Digest of Tech. Papers*, May 1997, pp. 285 – 292.

[60] D. Robertson, "Specifications and Figures of Merit for mixed-signal circuits," ISSCC Tutorial, 2002.

[61] B. Murmann, "Limits on ADC power dissipation," in *Workshop on Advances in Analog Circuit Design (AACD)*, Apr. 2005, accepted for publication.

[62] J. Voorman, "Transconductance amplifier," U.S. Patent4,723,110, Feb. 2, 1988.

[63] J. Voorman, W. Bruls, and P. Barth, "Bipolar integration of analog gyrator and Laguerre type filters," in *Proc. ECCTD*, Sept. 1983, pp. 108–110.

[64] B. Nauta, "A CMOS transconductance-C filter for very high frequencies," *IEEE J. Solid-State Circuits*, vol. 27, no. 2, pp. 142–153, Feb. 1992.

[65] B. Gilbert, "A high-performance monolithic multiplier using active feedback," *IEEE J. Solid-State Circuits*, vol. 9, no. 6, pp. 364–373, Dec. 1974.

[66] R. Crochiere and L. Rabiner, *Multirate digital signal processing.* Englewood Cliff, NJ: Prentice Hall, 1983.

[67] ——, "Interpolation and decimation of digital signals: a tutorial review," *Proc. IEEE*, vol. 69, no. 3, pp. 300–331, Mar. 1981.

[68] D. Goodman and M. Carey, “Nine digital filters for decimation and interpolation,” *IEEE Trans. Acoust., Speech, Signal Processing*, vol. 25, no. 2, pp. 121–126, Apr. 1977.

[69] J. Candy, “Decimation for sigma-delta modulation,” *IEEE Trans. Communications*, vol. 34, no. 1, pp. 72–76, Jan. 1986.

[70] E. Hogenauer, “An economical class of digital filters for decimation and interpolation,” *IEEE Trans. Acoust., Speech, Signal Processing*, vol. 29, no. 2, pp. 155–162, Apr. 1981.

[71] D. Tufts, D. Rorabacher, and W. Mosier, “Designing simple, effective digital filters,” *IEEE Trans. Audio Electroacoust.*, vol. 18, pp. 142–158, June 1970.

[72] H. Veendrick, *Deep-submicron CMOS ICs.* Kluwer Bedrijfsinformatie, 1998.

[73] D. Birru, “Reduced-sample-rate sigma-delta modulation,” Ph.D. dissertation, Delft Univ. of Techn., The Netherlands, June 1998.

[74] R. Schreier, “An empirical study of high-order single-bit delta-sigma modulators,” *IEEE Trans. Circuits and Systems II*, vol. 40, no. 8, pp. 461–467, Aug. 1993.

[75] B. Boser and B. Wooley, “The design of sigma-delta modulation analog-to-digital converters,” *IEEE J. Solid-State Circuits*, vol. 23, pp. 1298–1308, Dec. 1981.

[76] R. Adams, “Design and implementation of an audio 18-bit analog-to-digital converter using oversampling techniques,” in *J. Audio Eng. Soc.*, vol. 35, Mar. 1986, pp. 153–166.

[77] R. Adams, P. Ferguson, A. Ganesan, A. V. S. Vincelette, and R. Libert, “Theory and practical implementation of a fifth-order sigma-delta A/D converter,” in *J. Audio Eng. Soc.*, vol. 39, July 1991, pp. 515–528.

[78] J. Candy and O. Benjamin, “The structure of quantization noise from sigma-delta modulation,” *IEEE Trans. Communications*, vol. 29, no. 9, pp. 1316–1323, Sept. 1981.

[79] R. Gray, “Quantization noise spectra,” *IEEE Trans. Inform. Theory*, vol. 36, no. 6, pp. 1220–1244, Nov. 1990.

[80] D. Reefman and E. Janssen, “DC analysis of high order Sigma Delta modulators,” in *Proc. AES 113th Convention, paper 5693*, Sept. 2002.

[81] E. Roza, “Analog-to-digital conversion via duty-cycle modulation,” *IEEE Trans. Circuits and Systems II*, vol. 44, pp. 907–914, 1997.

[82] D. Reefman and E. Janssen, *Description of limit cycles in Sigma Delta Modulators.* PRLE Tech. Note 201028, Philips internal report, 2004.

[83] V. Friedman, “The structure of limit cycles in sigma-delta modulation,” *IEEE Trans. Communications*, vol. 36, no. 8, pp. 972–979, Aug. 1998.

[84] R. van Veldhoven, “A triple-mode continuous-time ΣΔ modulator with switched-capacitor feedback DAC for a GSM-EDGE/CDMA200/UMTS receiver,” *IEEE J. Solid-State Circuits*, vol. 38, no. 12, pp. 2069–2076, Dec. 2003.

[85] L. Breems, R. Rutten, and G. Wetzker, “A cascaded continuous-time ΣΔ modulator with 67-dB dynamic range in 10-MHz bandwidth,” *IEEE J. Solid-State Circuits*, vol. 39, no. 12, pp. 2152– 2160, Dec. 2004.

[86] DigRF: the digital interface standard. Digital Interface Working Group. [Online]. Available: www.digrf.com

[87] [Online]. Available: www.mipi.org

[88] JC−61 White Paper. JEDEC. [Online]. Available: www.jedec.org/Home/jc−61−Info/TechForum579 − JC − 61−Presentation.pdf

[89] G. den Besten, “Embedded low-cost 1.2Gb/s inter-IC serial data link in 0.35μm CMOS,” in *ISSCC Digest of Tech. Papers*, vol. 43, Feb. 2000, pp. 250–251.

[90] J. W. Whikehart, “DSP-based radio with IF processing,” presented at the SAE World Cong., Mar. 2000.

[91] K. Kianush and C. Vaucher, “A global car radio IC with inaudible signal quality checks,” in *ISSCC Digest of Tech. Papers*, Feb. 1998, pp. 130–131.

[92] E. Cherry and D. Hooper, *Amplifying devices and low-pass amplifier design.* New York: John Wiley and Sons, 1968.

[93] K. Philips and E. Dijkmans, “A variable gain IF amplifier with $-67dBc\ IM_3$-distortion at $1.4V_{pp}$ output in 0.25 μm CMOS,” in *VLSI Circuits Digest of Tech. Papers*, June 2001, pp. 81–82.

[94] J. Rijns, “CMOS low-distortion high-frequency variable-gain amplifier,” *IEEE J. Solid-State Circuits*, vol. 31, pp. 1029–1034, July 1996.

[95] C.-C. Hsu and J.-T. Wu, “A highly linear 125MHz CMOS switched-resistor programmable-gain amplifier,” *IEEE J. Solid-State Circuits*, vol. 38, no. 10, pp. 1663–1670, Oct. 2003.

[96] L. Vogt, D. Brookshire, S. Lottholz, and G. Zwiehoff, “A two-chip digital car radio,” in *ISSCC Digest of Tech. Papers*, Feb. 1996, pp. 350–351.

[97] S. Jantzi, W. Snelgrove, and P. Ferguson, “A fourth-order band-pass sigma-delta modulator,” *IEEE J. Solid-State Circuits*, vol. 28, pp. 282–291, Mar. 1993.

[98] F. W. Singor and W. M. Snelgrove, “Switched-capacitor bandpass delta sigma A/D modulation at 10.7MHz,” *IEEE J. Solid-State Circuits*, vol. 30, pp. 184–192, Mar. 1995.

[99] J. Park, E. Joe, M.-J. Choe, and B.-S. Song, “A 5MHz IF digital FM demodulator,” *IEEE J. Solid-State Circuits*, vol. 34, pp. 3–11, Jan. 1999.

[100] J. van Engelen, R. van de Plassche, E. Stikvoort, and A. Venes, "A sixth-order continuous-time bandpass sigma-delta modulator for digital radio IF," *IEEE J. Solid-State Circuits*, vol. 34, pp. 1753–1764, Dec. 1999.

[101] A. Ong and B. Wooley, "A two-path bandpass sigma-delta modulator for digital IF extraction at 10MHz," *IEEE J. Solid-State Circuits*, vol. 32, pp. 1920–1934, Dec. 1997.

[102] T. Ueno, A. Yasuda, T. Yamaji, and T. Itakura, "A fourth-order bandpass $\Delta\Sigma$ modulator using second-order bandpass noise-shaping dynamic element matching." *IEEE J. Solid-State Circuits*, vol. 37, no. 7, pp. 809–816, July 2002.

[103] I. Galton, W. Huff, P. Carbone, and E. Siragusa, "A delta-sigma PLL for 14-b 50-ksample/s frequency-to-digital conversion of a 10MHz FM signal," *IEEE J. Solid-State Circuits*, vol. 33, pp. 2042–2052, Dec. 1998.

[104] J. Crols and M. Steyaert, "Low IF topologies for high-performance analog front-ends of fully integrated receivers," *IEEE Trans. Circuits and Systems II*, vol. 45, no. 4, pp. 269–282, Mar. 1998.

[105] T. Lee, *The design of CMOS radio-frequency integrated circuits.* Cambridge, U.K.: Cambridge Univ.Press, 1998.

[106] J. Volder, "The CORDIC trigonometric computing technique," *IRE Trans. Electron. Comput.*, vol. EC-8, pp. 330–334, Sept. 1959.

[107] Q. Sandifort, L. Breems, C. Dijkmans, and H. Schuurmans, "IF-to-digital converter for FM/AM/IBOC radio," in *ESSCIRC Digest of Tech. Papers*, Sept. 2003, pp. 707 – 710.

[108] P. Quinn, K. van Hartingsveldt, and A. van Roermund, "A 10.7-MHz CMOS SC radio IF filter using orthogonal hardware modulation," *IEEE J. Solid-State Circuits*, vol. 35, no. 12, pp. 1865–1876, Dec. 2000.

[109] A. Nagari and G. Nicollini, "A 3v 10MHz pseudodifferential SC bandpass filter using gain enhancement replica amplifier," *IEEE J. Solid-State Circuits*, vol. 33, pp. 626–630, Apr. 1998.

[110] J. Silva-Martinez, M. Steyaert, and W. Sansen, "A 10.7MHz 68dB SNR CMOS continuous-time filter with on-chip automatic tuning," in *ISSCC Digest of Tech. Papers*, Feb. 1992, pp. 66–67.

[111] B.-S. Song, "A 10.7MHz switched-capacitor bandpass filter," *IEEE J. Solid-State Circuits*, vol. 24, pp. 320–324, Apr. 1989.

[112] Bluetooth standard. Bluetooth Special Interest Group. [Online]. Available: www.bluetooth.com

[113] S. Byun, C.-H. Park, Y. Song, S. Wang, C. Conroy, and B. Kim, "A low-power CMOS Bluetooth RF transceiver with a digital offset canceling DLL-based GFSK demodulator," *IEEE J. Solid-State Circuits*, vol. 38, no. 10, pp. 1609–1618, Oct. 2003.

[114] B. Xia, C. Xin, W. Sheng, A. Y. Valero-Lopez, and E. Sanchez-Sinencio, "A GFSK demodulator for low-IF Bluetooth receiver," *IEEE J. Solid-State Circuits*, vol. 38, no. 8, pp. 1397–1400, Aug. 2003.

[115] H. Darabi, S. Khorram, H.-M. Chien, M.-A. Pan, S. Wu, S. Moloudi, J. C. Leete, J. J. Rael, M. Syed, R. Lee, B. Ibrahim, M. Rofougaran, and A. Rofougaran, "A 2.4GHz CMOS transceiver for Bluetooth," *IEEE J. Solid-State Circuits*, vol. 36, no. 12, pp. 2016–2024, Dec. 2001.

[116] W. Sheng, B. Xia, A. E. Emira, C. Xin, A. Y. Valero-Lopez, S. T. Moon, and E. Sanchez-Sinencio, "A 3V, $0.35\mu m$ CMOS Bluetooth receiver IC," *IEEE J. Solid-State Circuits*, vol. 38, no. 1, pp. 30–42, Jan. 2003.

[117] P. Laurent, "Exact and approximate construction of digital phase modulations by superposition of Amplitude Modulated Pulses(AMP)," *IEEE Trans. Communications*, vol. COM-34, no. 2, pp. 150–160, Feb. 1986.

[118] N. Filiol, N. Birkett, J. Cherry, C. Cojocaru, A. Namdar, T. Pamir, K. Sheikh, G. Glandon, D. Payer, A. Swaminathan, R. Forbes, T. Rilley, S. M. Alinoor, E. Macrobbie, M. Clouter, S. Pipilos, and T. Varelas, "A 22mW Bluetooth RF transceiver with direct RF demodulation and on-chip IF filtering," in *ISSCC Digest of Tech. Papers*, Feb. 2001, pp. 202–203,447.

[119] M. Kokubo, M. Shida, T. Ishikawa, H. Sonoda, K. Yamamoto, T. Matsuura, M. Matsuoka, T. Endo, T. Kobayashi, K. Oosaki, T. Henmi, J. Kudoh, and H. Miyagawa, "A 2.4GHz RF transceiver with digital channel-selection filter for Bluetooth," in *ISSCC Digest of Tech. Papers*, Feb. 2002, pp. 94–95, 449.

[120] R. Staszewski, K. Muhammad, D. Leipold, C.-M. Hung, Y.-C. Ho, J. Wallberg, C. Fernando, K. Maggio, R. Staszewski, T. Jung, J. Koh, S. John, I. Deng, V. Sarda, O. Moreira-Tamayo, V. Mayega, R. Katz, O. Friedman, O. Eliezer, E. de Obaldia, and P. Balsara, "All-digital TX frequency synthesizer and discrete-time receiver for Bluetooth radio in 130-nm CMOS," *IEEE J. Solid-State Circuits*, vol. 39, no. 12, pp. 2278– 2291, Dec. 2004.

[121] H. Bergveld, K. van Kaam, D. Leenaerts, K. Philips, A. W. P. Vaassen, and G. Wetzker, "A low-power highly-digitized receiver for 2.4GHz-band GFSK applications," in *RFIC Digest of Tech. Papers*, June 2004, pp. 347 – 350.

[122] L. Eggermont, "Measured signal/quantising-distortion ratio of high-information delta modulation," *Electronics Letters*, vol. 11, no. 11, pp. 242–244, May 1975.

[123] F. Chen and S. R. B. Bakkaloglu, "A 1.5V 1mA 80dB passive ΣΔ ADC in 0.13μm digital CMOS process," in *ISSCC Digest of Tech. Papers*, vol. 1, Feb. 2003, pp. 54 – 55, 477.

[124] S. Yan and E. Sanchez-Sinencio, "A continuous-time sigma-delta modulator with 88-dB dynamic range and 1.1-MHz signal bandwidth," *IEEE J. Solid-State Circuits*, vol. 36, no. 12, pp. 75 – 86, Dec. 2001.

[125] J. Grilo, I. Galton, K. Wang, and R. Montemayor, "A 12-mW ADC delta-sigma modulator with 80dB of dynamic range integrated in a single-chip Bluetooth transceiver," *IEEE J. Solid-State Circuits*, vol. 37, no. 3, pp. 271–278, Mar. 2002.

[126] K. Doris, "High-speed D/A converters," Ph.D. dissertation, Eindhoven Univ. of Techn., The Netherlands, Sept. 2004.

[127] P. van Zeijl, J.-W. T. Eikenbroek, P. P. Vervoort, S. Setty, J. Tangenberg, G. Shipton, E. Kooistra, I. C. Keekstra, D. Belot, K. Visser, E. Bosma, and S. C. Blaakmeer, "A Bluetooth radio in $0.18\mu m$ CMOS," *IEEE J. Solid-State Circuits*, vol. 37, no. 12, pp. 1679–1687, Dec. 2002.

[128] E. Hioe, K. Maio, T. Oshima, Y. Shibahara, T. Doi, K. Ozaki, and S. Arayashiki, "0.18μm CMOS Bluetooth analog receriver with $-88dbm$ sensitivity," *IEEE J. Solid-State Circuits*, vol. 39, no. 2, pp. 374–377, Feb. 2004.

[129] O. Bajdechi and J. Huijsing, "A 1.8-V ΔΣ modulator interface for an electret microphone with on-chip reference," *IEEE J. Solid-State Circuits*, vol. 37, no. 3, pp. 279–285, Mar. 2002.

[130] J. Sauerbrey and R. Thewes, "Ultra low voltage switched opamp ΣΔ modulator for portable applications," in *CICC Digest of Tech. Papers*, May 2001, pp. 35–38.

[131] Y. Yang, A. Chokhawala, M. Alexander, J. Melanson, and D. Heser, "A 114-dB 68-mW chopper-stabilized stereo multibit audio ADC in $5.62mm^2$," *IEEE J. Solid-State Circuits*, vol. 38, no. 12, pp. 2061–2068, Dec. 2003.

[132] M. Keskin, U.-K. Moon, and G. Temes, "A 1-V 10-MHz clock-rate 13-bit CMOS ΔΣ modulator using unity-gain-reset opamps," *IEEE J. Solid-State Circuits*, vol. 37, no. 7, pp. 817–824, July 2002.

[133] E. Fogleman, J. Welz, and I. Galton, "An audio ADC ΔΣ -modulator with 100-dB peak SINAD and 102-dB DR using a second-order mismatch-shaping DAC," *IEEE J. Solid-State Circuits*, vol. 36, no. 3, pp. 339–348, Mar. 2001.

[134] K. Nguyen and K. Sweetland, "A 105dB SNR multibit ΣΔ ADC for digital audio applications," in *CICC Digest of Tech. Papers*, May 2001, pp. 27–30.

[135] M. Dessouky and A. Kaiser, "Very low-voltage digital-audio ΔΣ modulator with 88-dB dynamic range using local switch bootstrapping," *IEEE J. Solid-State Circuits*, vol. 36, no. 3, pp. 349 – 355, Mar. 2001.

[136] A. Dezzani and E. Andre, "A 1.2-V dual-mode WCDMA/GPRS ΣΔ modulator," in *ISSCC Digest of Tech. Papers*, vol. 1, Feb. 2003, pp. 58 – 59.

[137] H. Aboushady, F. Montaudon, F. Paillardet, and M. Louerat, "A 5mW, 100kHz bandwidth, current-mode continuous-time ΣΔ modulator with 84dB dynamic range," in *ESSCIRC Digest of Tech. Papers*, Sept. 2002, pp. 283–286.

[138] A. Nagari, A. Mecchia, E. Viani, S. Pernici, P. Confalonieri, and G. Nicollini, "A 2.7-V 11.8-mW baseband ADC with 72-dB dynamic range for GSM applications," *IEEE J. Solid-State Circuits*, vol. 35, no. 6, pp. 798 – 806, June 2000.

[139] O. Oliaei, P. Clement, and P. Gorisse, "A 5-mW ΣΔ modulator with 84-dB dynamic range for GSM/EDGE," *IEEE J. Solid-State Circuits*, vol. 37, no. 1, pp. 2–10, Jan. 2002.

[140] V. Cheung, H. Luong, and W.-H. Ki, "A 1-V 10.7-MHz switched-opamp bandpass ΣΔ modulator using double-sampling finite-gain-compensation technique," *IEEE J. Solid-State Circuits*, vol. 37, no. 10, pp. 1215–1225, Oct. 2002.

[141] P. Cusinato, D. Tonietto, F. Stefani, and A. Baschirotto, "A 3.3-V CMOS 10.7-MHz sixth-order bandpass ΣΔ modulator with 74-dB dynamic range," *IEEE J. Solid-State Circuits*, vol. 36, no. 4, pp. 629 – 638, Apr. 2001.

[142] M. Miller and G. Petrie, "A multi-bit ΣΔ ADC for multi-mode receivers," *IEEE J. Solid-State Circuits*, vol. 38, no. 3, pp. 475–482, Mar. 2003.

[143] T. Burger and Q. Huang, "A 13.5mW, 185MSample/s ΔΣ -modulator for UMTS/GSM dual-standard IF reception," *IEEE J. Solid-State Circuits*, vol. 36, no. 12, pp. 1868–1878, Dec. 2001.

[144] P. Rombouts, , J. D. Maeyer, and L. Weyten, "A 250kHz 94dB double-sampling ΣΔ modulation A/D converter with a modified noise transfer function," *IEEE J. Solid-State Circuits*, vol. 38, no. 10, pp. 1657–1662, Oct. 2003.

[145] P. Maulik, M. Chadha, W. Lee, and P. Crawley, "A 16-bit 250-kHz delta-sigma modulator and decimation filter," *IEEE J. Solid-State Circuits*, vol. 35, no. 4, pp. 458 – 467, Apr. 2000.

[146] F. Esfahani, P. Basedau, R. Ryter, and R. Becker, "A fourth order continuous-time complex sigma-delta ADC for low-IF GSM and EDGE receivers," in *VLSI Circuits Digest of Tech. Papers*, June 2003, pp. 75 – 78.

[147] T. Salo, S. Lindfors, T. Hollman, J. Jarvinen, and K. Halonen, "80MHz bandpass ΔΣ modulators for multimode digital IF receivers," *IEEE J. Solid-State Circuits*, vol. 38, no. 3, pp. 464–474, Mar. 2003.

[148] T. Salo, S. Lindfors, and K. Halonen, "A 80MHz bandpass ΔΣ modulator for a 100-MHz IF receiver," *IEEE J. Solid-State Circuits*, vol. 37, no. 7, pp. 798–808, July 2002.

[149] R. Schreier, J. L. ana L. Singer, D. Paterson, M. Timko, M. Hensley, G. Patterson, K. Behel, and J. Zhou, "A 10-300-MHz IF-digitizing IC with 90-105-dB dynamic range and 15-333-kHz bandwidth," *IEEE J. Solid-State Circuits*, vol. 37, no. 12, pp. 1636–1644, Dec. 2002.

[150] R. Gaggl, A. Wiesbauer, G. Fritz, C. Schranz, and P. Pessl, "A 85-dB dynamic range multibit delta-sigma ADC CMOS bandpass continuous-time multibit Δ-Σ ADC for ADSL-CO applications in 0.18μm CMOS," *IEEE J. Solid-State Circuits*, vol. 38, no. 7, pp. 1105–1114, July 2003.

[151] K. Philips, P. Nuijten, R. Roovers, A. van Roermund, F. M. noz, M. T. Pallares, and A. Torralba, "A continuous-time $\Sigma\Delta$ ADC with increased immunity to interferers," *IEEE J. Solid-State Circuits*, vol. 39, no. 12, pp. 2170– 2178, Dec. 2004.

[152] S. Gupta and V. Fong, "A 64-MHz clock-rate $\Sigma\Delta$ ADC with 88-dB SNDR and -105-dB IM_3 distortion at a 1.5-MHz signal frequency," *IEEE J. Solid-State Circuits*, vol. 37, no. 12, pp. 1653–1661, Dec. 2002.

[153] T.-H. Kuo, K.-D. Chen, and H.-R. Yeng, "A wideband CMOS sigma-delta modulator with incremental data weighted averaging," *IEEE J. Solid-State Circuits*, vol. 37, no. 1, pp. 11–17, Jan. 2002.

[154] I. Fujimori, L. Longo, A. Hairapetian, K. Seiyama, S. Kosic, C. Jun, and C. Shu-Lap, "A 90-dB SNR 2.5-MHz output-rate ADC using cascaded multibit delta-sigma modulation at 8 oversampling ratio," *IEEE J. Solid-State Circuits*, vol. 35, no. 12, pp. 1820 – 1828, Dec. 2000.

[155] T. Ueno and T. Itakura, "A 0.9V 1.5mW continuous-time $\Delta\Sigma$ modulator for WCDMA," in *ISSCC Digest of Tech. Papers*, Feb. 2004, pp. 78–79, 514.

[156] L. Dorrer, F. Kuttner, A. Wiesbauer, A. D. Giandomenico, and T. Hartig, "10-bit, 3mW continuous-time $\Sigma\Delta$ ADC for UMTS in a 0.12μm CMOS process," in *ESSCIRC Digest of Tech. Papers*, Sept. 2003, pp. 245–248.

[157] R. Jiang and T. Fiez, "A 1.8 V 14 b Δ-Σ A/D converter with 4MSamples/s conversion," in *ISSCC Digest of Tech. Papers*, Feb. 2002, pp. 220 – 461.

[158] K. Vleugels, S. Rabii, and B. Wooley, "A 2.5-V sigma-delta modulator for broadband communications applications," *IEEE J. Solid-State Circuits*, vol. 36, no. 12, pp. 1887–1899, Dec. 2001.

[159] A. Tabatabaei and B. A. Wooley, "A two-path bandpass SigmaDelta modulator with extended noise shaping," *IEEE J. Solid-State Circuits*, vol. 35, pp. 1799 – 1809, Apr. 2002.

[160] F. Ying and F. Maloberti, "A mirror image free two-path bandpass $\Sigma\Delta$ modulator with 72dB SNR and 86dB SFDR," in *ISSCC Digest of Tech. Papers*, Feb. 2004, pp. 84–85, 514.

[161] L. Luh, J. C. Jr., and J. Dapper, "A 400MHz 5th-order CMOS continuous-time switched current ΣΔ modulator," in *ESSCIRC Digest of Tech. Papers*, sep 2000, pp. 72–75.

[162] Y. Geerts, M. Steyaert, and W. Sansen, "A 12-bit 12.5 MS/s multi-bit ΔΣ CMOS ADC," in *CICC Digest of Tech. Papers*, May 200, pp. 21 – 24.

[163] P. Balmelli and Q. Huang, "A 25MS/s 14b 200mW ΣΔ modulatorin 0.18μm CMOS," in *ISSCC Digest of Tech. Papers*, Feb. 2004, pp. 74–75, 514.

[164] P. Wambacq and W. Sansen, *Distortion analysis of analog integrated circuits*. Norwell, MA: Kluwer Academic Publishers, 1998, pp. 375–384.

[165] H. Samueli, "A low-complexity multiplierless half-band recursive digital filter design," *IEEE Trans. Acoust., Speech, Signal Processing*, vol. 37, no. 3, pp. 442–444, Mar. 1989.